33rd Annual Saudi-Japan Symposium on Technology in Petroleum Refining & Petrochemicals 2025

Technology in Fuels & Petrochemicals

Dhahran, Saudi Arabia
10 - 11 February 2025

ISBN: 979-8-3313-1449-1

Printed from e-media with permission by:

Curran Associates, Inc.
57 Morehouse Lane
Red Hook, NY 12571

Some format issues inherent in the e-media version may also appear in this print version.

Printed with permission by Curran Associates, Inc. (2025)

For permission requests, please contact King Fahd University of Petroleum & Minerals at the address below.

King Fahd University of Petroleum & Minerals
Dhahran 31261
Kingdom of Saudi Arabia

Phone: +966 (13) 860 0000

info@kfupm.edu.sa

Additional copies of this publication are available from:

Curran Associates, Inc.
57 Morehouse Lane
Red Hook, NY 12571 USA
Phone: 845-758-0400
Fax: 845-758-2633
Email: curran@proceedings.com
Web: www.proceedings.com

الندوة السنوية السعودية اليابانية الثالثة والثلاثون

33rd Annual Saudi-Japan Symposium

التقنية في الوقود والبتروكيماويات

Technology in Fuels & Petrochemicals

ABSTRACTS: 33

12-11 شعبان 1446هـ

February 10-11, 2025

King Fahd University of Petroleum & Minerals

Dhahran, Saudi Arabia

Preface

This is the "Abstracts" of papers presented at the 33rd Annual Saudi Japan symposium "*Technology in Fuels & Petrochemicals*" held on February 10-11, 2025, at King Fahd University of Petroleum & Minerals (KFUPM), Dhahran, Saudi Arabia. The event, jointly organized by KFUPM Center for Refining & Advanced Chemicals, Saudi Aramco Catalyst Center of Excellence, Japan Cooperation Center for Petroleum and Sustainable Energy (JCCP), and the Japan Petroleum Institute (JPI), brings together leading researchers, scientists, engineers and industry professionals to explore breakthroughs and transformative advancements shaping the future of chemistry fuels and chemicals.

The main theme of this year's event is "*Advancing Sustainable Solutions*" with topics including liquids to chemicals, waste to chemicals, gas to liquids, alternative and renewable fuels and cheminformatics. Four insightful keynote speeches by distinguished speakers from KFUPM, KAUST, Hokkaido and Waseda universities. In addition, sixteen oral presentations by excellent panel of speakers from KFUPM, Saudi Aramco, OSP, JPI and Japanese universities of Tohoku and Tottori. In addition, there are 30 poster presentations by graduate students, postdocs and young researchers from Aramco and KFUPM.

I take this opportunity to thank the President of KFUPM Dr. Muhammad M. Al-Saggaf and the Vice President for Research and Innovation Dr. Ali A. Al-Shaikhi for their continuous support and valuable guidance for holding this Symposium. I would like to thank our counterparts at Saudi Aramco, JCCP and JPI for their efforts throughout the planning stages of the symposium. I also thank all members of organizing committee for preparing this event.

We extend our thanks to distinguished speakers and attendees for their active participation and look forward to your participation in our next symposium.

Dr. Wael Ahmed Fouad
Director, CRAC & PCRC
Associate Professor, Chemical Engineering Department
King Fahd University of Petroleum & Minerals
Dhahran, Saudi Arabia
https://ri.kfupm.edu.sa/crac

Contents

Posters

Oral 1 (Keynote)
Waste to Sustainable Aromatics, Olefins and Aviation Fuels by Zeolite-Based Catalysis

Konstantinos Triantafyllidis
Chemistry Department, King Fahd University of Petroleum and Minerals, Dhahran, Saudi Arabia
Interdisciplinary Research Center for Refining and Advanced Chemicals, King Fahd University of Petroleum & Minerals, Dhahran 31261, Saudi Arabia
E-mail: k.triantafyllidis@kfupm.edu.sa

Classical petroleum refineries can contribute to sustainable development by adopting alternative feedstocks and continuously improving the efficiency and environmental friendliness of downstream refining processes and catalysts. Valorizing waste streams like agro-industrial and municipal waste (e.g., used lipids and plastics) offers feedstock flexibility through (co)processing of derived crude oils (pyrolysis oils, hydrogenolysis oils, etc.), at the same time promoting the circular economy model. In conventional refineries, the transformation of heavy petroleum fractions into lighter hydrocarbon products (LPGs, gasoline, diesel, naphtha, kerosene, etc.) relies heavily on catalytic processes, many of which utilize zeolite-based catalysts. Zeolite ZSM-5 exemplifies the importance of this material class. Its unique three-dimensional, medium-sized (~5.5 Å) tubular micropore structure and enhanced Brønsted acidity provide tuned shape-selectivity and reactivity for various processes, perhaps the most important being Fluid Catalytic Cracking (FCC) where ZSM-5 complements the activity of the main USY zeolite catalyst component. ZSM-5 zeolite, as a cracking catalyst or transition/noble metal support, may also be very effective in the transformation of biomass, lipids and plastic waste into suitable crudes for co-processing or even high-quality chemicals and fuel components. Furthermore, ZSM-5 or its variants are promising catalysts for the Methanol-to-Gasoline/Olefins/Aromatics (MTG/O/A) process, crucial for sustainable CO_2 utilization. While ZSM-5 is prominent, other zeolites like Beta and Ferrierite, with different pore topologies, sizes, and Brønsted/Lewis acid site functionalities, are also effective in various waste/bio-refining processes. Hierarchical micro/meso/macro-porous zeolite formulations can further enhance performance, particularly where diffusion limitations are a concern.

This presentation will showcase and discuss recent applications of zeolitic catalysts in pyrolysis and hydrotreatment (hydro-deoxygenation/isomerization/cracking) of various waste streams towards the production of aromatics (including phenolics), light olefins and fuels.

Oral 2
Catalytic Transformation of LDPE into Fuel Oil Enriched with Aromatic Hydrocarbons

Muhammad Naseem Akthar
Interdisciplinary Research Center for Refining and Advanced Chemicals, King Fahd University of Petroleum and Minerals, Dhahran 31261, Saudi Arabia.
Department of Chemistry, King Fahd University of Petroleum & Minerals, Dhahran 31261, Saudi Arabia.

Plastics are non-biodegradable materials that are typically manufactured from non-renewable petroleum resources. Due to their high utilization in both industrial and domestic sectors, the production rate of plastics is significant. It has been reported that the generation of plastic has doubled in the last twenty years. Most of this plastic waste is either dumped into the landfills or incinerated causing pollution to the environment. Almost 90% of the plastic products are used only once and then discarded. The catalytic transformation of low-density polyethylene (LDPE) into aromatic-rich fuel oil is a promising approach to recycle plastic waste and produce valuable hydrocarbons. This process is typically performed via catalytic pyrolysis, where LDPE is broken down under heat in the presence of catalysts to produce aromatic hydrocarbons such as benzene, toluene, and xylenes (BTX).

The current study focuses on the conversion of LDPE into high-grade fuel oil using a catalytic degradation method. A semi-batch reactor was used for this purpose at 350 °C under ambient pressure and a catalyst to LDPE ratio of 1:20. Several zeolite-based catalysts were prepared by impregnation of different metals (Fe, Zn, Cr, Mn, and Ga) on ZSM-5 with a silica to alumina ratio of 30 (denoted as Z30). These catalysts were analyzed using BET, SEM, XRD, and NH_3-TPD, instrumental techniques to assess their physico-chemical properties. The results indicated that pyrolytic oil yield was dropped during catalytic pyrolysis of LDPE with respect to that of non-catalytic pyrolysis. The pyrolytic oil was characterized by using various analytical techniques, including elemental analysis, gross calorific value (GCV), Simulated Distillation, and GC-DHA. The elemental analysis revealed a high carbon (85-86%) and hydrogen (13-14%) content in the pyrolytic oil, contributing to a high GCV of ~42 MJ/kg. The GC-DHA analysis revealed that the pyrolytic oil was enriched with aromatic and olefinic compounds. The 5Fe/Z30 exhibited the highest aromatic selectivity (34%) and research octane number (91) along with 100% conversion. The analysis of pyrolytic oil suggests that catalytic pyrolysis of LDPE holds potential applications in the synthesis of valuable chemicals and fuels. The 5Fe/Z30 and catalysts were further evaluated using *ex-situ* pyrolysis process. In this case vapors of pyrolysis products were passed through a fixed bed reactor having 2Ga/Z30 aromatization catalyst at 550 °C The pyrolytic oil exhibited the highest aromatic selectivity of 95.5% under these reaction conditions.

Oral 3
Pyrolysis Oil Upgrading to Aromatics

Miao Sun
Fuels & Chemicals Division, R&D Center, Saudi Aramco, Dhahran 31311, Saudi Arabia

Background and motivation: In traditional refineries, heavy pyrolysis oil is produced from naphtha or mixed feed cracker as by-product, and in most cases sold as marine fuel oil or heating fuel oil. However, this will soon discontinue, due to the concerns on fast climate change and serious pollution caused by heavy fuel oil burning. In 2021, International Maritime Organization (IMO)'s introduced a prohibition on the use and carriage for use as fuel of heavy fuel oil (HFO) by ships in Arctic waters on and after July 1st, 2024.[1] The traditional demand & market of heavy pyrolysis oil, as marine fuel oil or heating fuel oil, will slump in the coming a few years.

As one of the world's largest integrated energy and chemicals producers, Aramco is always seeking ways to balance the needs of meeting global energy demand, and at the same time improving environmental performance in order to achieve the goal of net-zero GHGs emission (Scope 1 and Scope 2) by year 2050. Aramco's strategic focus is on developing and deploying innovative solutions, optimizing operations, and adopting efficient project designs that aim to mitigate GHGs emissions. As part of the wider materials transition, petrochemicals are set to be the primary engine of oil demand growth in the coming years, and Aramco is investing heavily in this sector to create and deploy technologies to produce highly demanded chemicals with less emissions.[2]

Finding green solutions for future refineries in order to continue creating values from low value liquids, the valorization of intermediate petrochemicals/refinery streams to petrochemical building blocks has attracted more and more attention. In conjunction, the market demand for aromatics, which serve as basic building blocks and intermediates in the petrochemical industry, has been steadily growing and the market price for these petrochemicals has shown tangible appreciation in the past decade. This trend creates an opportunity and a potential market for technologies that produce aromatic building blocks from cheap industrial intermediate streams.

Materials and methods: Saudi Aramco developed Pyrolysis Oil Upgrading to Aromatics (P2A) technology is a novel process that converts low value steam cracker pyrolysis oil (IBP 170 °C – FBP 732 °C) into high value mono-aromatic rich pyrolysis gasoline by using a multiple-stage process with separate fixed-bed reactors in series. This technology capitalizes on the rich aromatic and low sulfur content of pyrolysis oil to provide a more economical and environmentally friendly alternative to the current usage as fuel. P2A technology can valorize pyrolysis oil to a higher value mono-aromatics stream, increase the complex's pyrolysis gasoline yield and address the supply-demand gap of key petrochemicals. P2A technology aims at valorizing heavy pyrolysis oil to create more values, and bringing clean & green solutions to future refineries. P2A technology is a state-of-the-art technology, as no commercial competing technologies currently exist. Strong market demand and attractive financials make it economically attractive.

Results and discussion: The overall net yield of mono-aromatic products is more than 50wt%, while the light gases (C1-C4 gases) emission is as low as below 1wt%.

References:

[1] Amendments to MARPOL Annex I _ MEPC.329(76).

[2] Aramco Sustainability Report 2023.

Oral 4
Recycling of Waste Plastics Without Consumption of other Materials: Shape Selective Cracking of Polyolefin on H-MFI Zeolite Catalyst with Recovery of Cyclooctane Solvent

Naonobu Katada
Center for Research on Green Sustainable Chemistry, Faculty of Engineering, Tottori University
4-101 Koyama-cho Minami, Tottori 680-8552, Japan
Tel & Fax: +81-857-31-5684, E-mail: katada@tottori-u.ac.jp

The use of hydrocarbon solvents for zeolite-catalyzed polyolefin cracking narrows the molecular weight distribution of the products, which enhances the efficiency of polyolefin chemical recycling to naphtha, a key precursor to polyolefins. However, solvent consumption presents a challenge. In this study, zeolite microporosity was used to achieve shape-selective polyolefin cracking while allowing solvent recovery. With an H-MFI type zeolite catalyst combined with cyclooctane as the solvent, polypropylene was selectively converted without cyclooctane reactivity. In a typical case, 84% of polypropylene was converted into C3–27 aliphatic and monocyclic aromatic compounds (equivalent to liquid petroleum gas, naphtha, kerosene, jet and diesel) with 79% selectivity, while 95% of cyclooctane was recovered. This study is the first to demonstrate solvent recyclability in polyolefin cracking on an acidic zeolite, contribute to the chemical recycling of polyolefin into its precursor, naphtha, under high selectivity given by the presence of solvent but without solvent consumption.

In addition, the location of polyolefin cracking was analyzed from the activity change by alkali treatments of the MFI zeolite. Under the adopted conditions, the alkali treatment increased the number of Brønsted acid sites exposed on the external surface by dissolution of amorphous-like phase on the external surface of zeolite. Then, the external surface area was increased by formation of mesopores. Increase of the acid sites on the external surface and the external surface area by the alkali treatment have been known well, but the increases of them with the different treatment extents have been observed for the first time. On this series of zeolites, the polyolefin cracking activity was enhanced with increasing the external surface area but not with increasing the Brønsted acid sites on the external surface. It indicates that the major reaction field was not the Brønsted acid sites on the external surface, but the region close to the entrances inside micropores. This enables us to apply the shape selectivity as above.

Oral 5
Kinetic Influence of Ba on Cobalt-catalyzed Ammonia Decomposition Reaction

Dr. Zahra Almisbaa, and Philippe Sautet
Saudi Aramco
Department of Chemical and Biomolecular Engineering, University of California, Los Angeles, Los Angeles, California 90095, United States

Ammonia decomposition is a key reaction in the context of hydrogen storage, transport, and release. This study combines density functional theory (DFT) calculations with microkinetic modeling to address the promotion mechanism of Ba species for ammonia decomposition on Co catalysts. The modified adsorption properties of Co upon the addition of metallic Ba or BaO suggest that the promoters play a role in alleviating the competitive adsorption of H. Calculating the full reaction pathway of ammonia decomposition shows that limiting the investigation to the N–N association step, as done previously, overlooks the effect of the promoter on the energy barriers of the NHx dehydrogenation steps. Challenges of modeling the ammonia decomposition reaction are addressed by understanding that the NH2 intermediate is stabilized on the step sites rather than the terrace sites. When the effect of H-coverage on the adsorption of NH3 is not considered in the microkinetic simulations, the results conflict with the experiments. However, accounting for the effect of H-coverage, as performed here, shows that BaO-doped Co has higher rates than pristine Co and Ba-doped Co at the reaction temperature of 723.15 K. When H is adsorbed on the Ba-doped Co, the adsorption of ammonia becomes significantly endergonic, which makes the rates relatively slow. The superiority of the BaO-promoted catalyst is attributed to a lower energy for the transition state of the rate-determining step, coupled with a reduced impact of the hydrogen coverage on weakening the ammonia adsorption. The kinetic analysis of the influence of Ba and BaO on the Co surface shows that BaO-doped Co aligns more closely with experimental observations than Ba-doped Co. This implies that Ba on the Co surface is likely to be in an oxide form under reaction conditions. Understanding the kinetics of the ammonia decomposition reaction provides a foundation for developing highly effective catalysts to accelerate the industrial utilization of ammonia as a sustainable hydrogen carrier.

Oral 6
Sustainable Passive Solutions for Reduced Energy Use in Cooling Applications

Issam Gereige, Konstantinos Kotsovos, Ali Basaheeh, Sarah Aqeel
Saudi Aramco Research and Development Center, KAUST R&D Center, Thuwal, Saudi Arabia

Heat stress is one of the most serious climate threats to dense communities, which is further exacerbated in recent years due to climate warming, and it accounts for 20% of electricity consumption worldwide. Therefore, employing reliable and energy-efficient cooling techniques significantly contributes to addressing this challenge and enhances sustainability. Here, we showcase a bifunctional cooling technique with passive radiative and evaporative functionalities by an easy-to-fabricate and cost-effective hygroscopic sodium polyacrylate (PAAS) hydrogel film. The PAAS photonic film with high solar reflectance (95%) and mid-infrared emittance (99%) offers high-performance daytime radiative cooling, while it absorbs atmospheric water vapor for higher cooling performance. The nocturnal radiative cooling of the PAAS photonic film resulting from its high mid-infrared emittance accelerates vapor condensation during the nighttime and enables evaporative cooling for the daytime, exemplified by a 5-8°C temperature reduction under a high solar irradiance of $900 W.m^{-2}$. PAAS photonic film can be fabricated using roll-to-roll technique, enabling high-throughput and economical fabrication. The cost-effective raw material and scalable fabrication ensure its broad applicability to building envelopes, as well as other sectors such as hydrogen tanks and energy storage solutions.

Oral 7
Sustainability Without Borders: Advancing Partnerships Between the Kingdom of Saudi Arabia and Japan

Zainab Nasif and Ahaad Turki
Oil Sustainability Program, OSP, Riyadh, Saudi Arabia
Emails: zainab.nasif@osp.gov.sa ahaad.turki@osp.gov.sa

The pursuit of sustainability transcends borders, fostering collaborative local and international relationships drive sustainable innovation and shared progress. Through this presentation, OSP aims to highlight the role we play on both the private and public levels, how one cannot function without the other. We aim to exemplify the Kingdom and Japan's commitment towards sustainability, but most importantly, to clearly show our commitment to expand our global partnerships with Japan even further. We will provide an overview of our Program, emphasizing key opportunities for cooperation and showcasing a recent successful initiative that exemplifies the impact of Saudi-Japanese collaboration. By leveraging expertise, resources, and joint efforts, we aim to advance sustainable solutions that contribute to global environmental and economic resilience in the face of uncertainties and challenges.

Oral 8
North American Market Research – Energy Transition to Renewables from Policies and Technologies

Satoshi Ogano
Japan Petroleum Institute
1-8-4 Kanda-Sudacho, Chiyoda-ku, Tokyo 101-0041, Japan
Tel: +81-3-6206-4301, E-mail: satoshi.ogano@sekiyu-gakkai.or.jp

Energy transition in North America would be a good mirror of our future, because advanced policies have been already established. One of the examples would be an increased EV implementation policy to reduce CO_2 emission reduction in the US. From my perspective, however, this would not be going well so far, since the target of EV implementation ratio would be extremely high in 2032, based on EPA's regulation[1], and thus the other CO_2 emission reduction options need to be focused instead.

From carbon neutral standpoints, there are lots of areas of the technologies developments needed from oil industries, especially refiners. North America is one of the biggest oil producing areas in the world and thus what is happening in North America for carbon neutral looks like a good reference for the Middle East.

Feedstock and refining integration would be one of the good approaches to achieve our carbon neutral society without too much sacrificing our ongoing economic activities. Several examples will also be shown to demonstrate what is going on around refineries and their feedstocks.

In addition, US refineries can provide a lot of data to understand the current global energy status and the background of why this happens. This is well reflecting in the petroleum product price, supply capability and product stocks, which are linked with each other to respond to the market demand.

Reference:

[1] US Federal Register Vol.89, No.76, Apr 18, 2024, pp.27842-28211

Oral 9
Ultrasonically Enhanced Oxidative Desulfurization of Heavy Fuel Oil – Pilot Scale Test Results and Techno-Economics Assessment

Shekhar R Kulkarni[1], Musaed Ghrami[2], Deoras Prabhudharwadkar[1], Dossary, Mohammed[2], Paolo Guida[1], Elia Colleoni[1], Varaha Prasad Sarvothaman[1], William Roberts[1]

[1]*King Abdullah University of Science and Technology, Thuwal, Makkah, 23955, Saudi Arabia.*

[2]*Saudi Aramco Company, Dhahran, Saudi Arabia*

A new limit on the sulfur content in the fuel oil used onboard ships came into force in 2020 (known as IMO 2020), marking a significant milestone to improve air quality, preserve the environment and protect human health. This regulation limits the sulfur in the fuel oil for the ships to 0.50% (by mass), which is a significant reduction from the previous limit of 3.5%. The most commonly used fuel for marine industry is the heavy fuel oil (HFO) with sulfur percentages ~3.5%. The bunker price difference between the heavy fuel oil and very low sulfur fuel oil (VLSFO, with sulfur mass of 0.5%) is about 140-180 USD/metric ton and has a strong sensitivity to the geopolitical and market dynamics. This has been a primary motivation to investigate various desulfurization technologies. While carbon-free fuels are being developed as a long-term solution for decarbonizing the maritime emissions, affordable VLSFO will be key to sustainable solution during the transition.

Oxidative Desulfurization (ODS) process is a widely investigated for removing sulphur-containing compounds, such as thiophenes and sulfides, from both light and heavy petroleum fractions. This method involves blending an oxidizing agent with the hydrocarbon mixture, typically in the presence of an acid catalyst. The oxidizing agent converts sulphur compounds into sulfones. These species are more polar than the parent thiophene, and thus can be extracted efficiently by a solvent yielding VLSFO. Despite its straightforward concept, the conventional ODS processes have been economically impractical due to the high consumption of oxidizing agents and prolonged reaction times, resulting in low process efficiency and throughput.

A novel continuous Ultrasonically Induced Cavitation (UIC) assisted oxidation technology developed at KAUST effectively overcomes the limitations of traditional methods, making the process economically feasible. UIC employs a vibrating sonotrode to generate pressure waves, causing the formation of nanoscale bubbles in the reaction medium. These bubbles rapidly nucleate, oscillate, and collapse within a much shorter timescale than the surrounding flow field. This dynamic action breaks down the asphaltene aggregates, increasing the exposure of sulphur molecules to the oxidizing agent. Additionally, UIC significantly enhances the surface area between oxidizer droplets and oil, drastically reducing the oxidizer consumption.

This paper presents an overview of the ultrasonically enhanced oxidative desulfurization (uODS) process, updates on its current development stage, and a techno-economic analysis of a uODS system. Also, the validation of proof-of-concept will be highlighted.

Oral 10
Catalysis Reinvented: Sustainable Materials and Ultrafast Laser Insights for Economic And Environmental Advancement

Ahmed M. El-Zohry
Ultrafast Laser Spectroscopy Lab, Center for Integrative Petroleum Research, KFUPM, Dhahran 31261, Saudi Arabia
Email: ahmed.elzohry@kfupm.edu.sa

Conversion of low-value materials, particularly from crude oil, into high-value products is a critical process for industrial and economic applications. Traditionally, this transformation relies on non-renewable and costly resources, such as precious metals (e.g., platinum and gold) or toxic materials based on cadmium- and lead-based compounds. Addressing this challenge from an environmental perspective, we focus on employing sustainable materials capable of efficiently harnessing sunlight to drive a wide range of chemical reactions, aiming for economically viable outcomes. To unravel the catalytic mechanisms of these sustainable materials, we employ ultrafast laser pulses on femtosecond timescales across different energy ranges, allowing us to probe both the electronic and vibrational dynamics of the catalysts and organic substrates. This multidimensional approach provides a comprehensive understanding of the reaction pathways, shedding light on how sustainable catalysts facilitate the conversion of low-value feedstocks into high-value products. In this talk, we will showcase several successful examples of organic transformations catalyzed by transition-metal-based sustainable materials. These reactions, driven by sunlight and monitored through ultrafast spectroscopic techniques, demonstrate the potential to reduce reliance on costly and harmful materials while achieving efficient and selective chemical processes. Our findings not only reveal new insights into the catalytic behavior of these materials but also highlight their practical applications in accelerating economic growth through

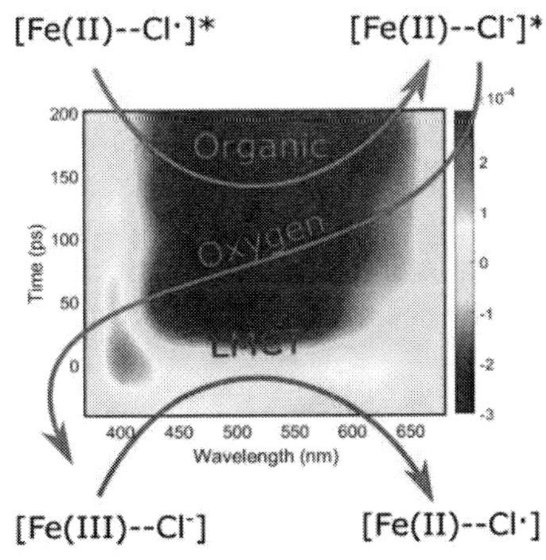

Scheme 1: Graphical Illustration for the excited state dynamics for one of the sustainable catalysts for oxidizing organic substrates

Oral 11 (Keynote)
Accelerated Discovery of Heterogeneous Catalysts using Machine Learning

Takashi Toyao
Institute for Catalysis, Hokkaido University
N-21, W-10, Sapporo 001-0021, Japan
Tel & Fax: +81 -11-706-9165, E-mail: toyao@cat.hokudai.ac.jp

Recent revolution in data science could accelerate the pace of new catalyst development. Machine learning (ML), a sub-field of data science, would play a central role in this paradigm shift. Despite having already seen proof-of-concept examples that ML can reduce the time and costs involved, most research has been directed toward the resolution of benchmark problems, but truly novel compounds and materials have rarely been discovered. This is due to one of the most common limitations of ML—an assumed inability for models to extrapolate, i.e. to identify extraordinary materials beyond those present in the training data set. In materials and catalysis informatics, we often desire to use ML models to find entirely new classes of materials and catalysts, with heretofore-unseen combinations of elements.

In this context, our group has developed a new ML approach that uses elemental features as the input representations, rather than inputting the catalyst compositions directly.[1,2] In our method, the elemental composition ratios are multiplied by elemental descriptors such as electronegativity and atomic radius, which are unique for each element. This ML method has the potential to guide catalyst design and discovery in areas where limited catalyst composition overlap exists, even for elements previously unseen in the given dataset. This will enable us to make extrapolative prediction beyond the training data.

In this study, we have applied the extrapolative ML approach to develop new multi-elemental catalysts based on supported Pt as an active metal and TiO_2 as a support for the low-temperature reverse water-gas shift (RWGS) reaction.[3] Using 45 catalysts as the initial data points and performing 44 cycles of the closed loop discovery system (ML prediction + experiment), we experimentally tested a total of 300 catalysts and identified more than 100 catalysts with superior activity compared to those of the previously reported high-performance catalysts. The composition of the optimal catalyst discovered was Pt(3)/Rb(1)-Ba(1)-Mo(0.6)-Nb(0.2)/TiO_2. Notably, niobium (Nb) was not included in the original dataset, and the catalyst composition identified was not predictable even by human experts. In my talk, I will also present some recent works on other catalytic reactions.

Reference:

[1] S. Mine, M. Takao, T. Yamaguchi, T. Toyao, Z. Maeno, S. M. A. Hakim Siddiki, S. Takakusagi, K. Shimizu and I. Takigawa, *ChemCatChem*, **2021**, 13, 3636–3655.

[2] S. Mine, Y. Jing, T. Mukaiyama, M. Takao, Z. Maeno, K. Shimizu, I. Takigawa and T. Toyao, *Chem. Lett.*, **2022**, 51, 269–273.

[3] G. Wang, S. Mine, D. Chen, Yuan Jing, K. W. Ting, T. Yamaguchi, M. Takao, Z. Maeno, I. Takigawa, K. Matsushita, K. Shimizu and T. Toyao, *Nature Communications*, **2023**, 14, 5861.

Oral 12
Mechanochemical Route to Synthesize Zeolites as Novel Method and their Characterization with XAFS

Atsushi Muramatsu
Graduate School of Agricultural Science, Center for Agricultural and Life Sciences using Synchrotron Light (A-Sync), Tohoku University
Aramaki, Aoba-ku, Sendai, 980-8572, Japan
E-mail: mura@tohoku.ac.jp

Novel synthesis method for zeolite materials has been developed in our laboratory, which is based on mechanochemical reaction (MC) of precursor solid oxides, containing amorphous SiO_2 and the other metal (hydro)oxides, to give completely amorphous mixed oxides, and successive hydrothermal synthesis through conversion from them via solution phase. Ce-containing MFI-type zeolite (Ce-MFI)-supported Pd catalysts modified by a second metal were prepared and found active for low-temperature oxidative coupling of methane reactions to give C_2 hydrocarbons. The Ce species in this sample were more highly dispersed than those in the two control materials: A Ce-containing MFI-type zeolite synthesized via the conventional one-pot hydrothermal approach and Ce species simply deposited on silicalite-1. Separately, Fe-MWW-type zeolite was synthesized using a MC reaction. The change in the Fe local structure due to the MC reaction was tracked from the Fe K-edge XAFS spectrum. It was found that the Fe-Fe bonds derived from iron oxide decreased and the coordination number decreased depending on the MC treatment time. This suggests that the a-FeOOH used as the raw material was dissolved in SiO_2 as FeO_4 by the MC reaction, and the degree of substitution changed with the MC treatment time.

Oral 13
Catalytic Cracking Technology for VGO Application

Maryam M Al-Taher, Prabha Dasila, Ali Jishi
R&DC, Saudi Aramco
Kido Naoyuki, Mizuno Takaki, Takamiya Yusuke, Ishihara Hisaya
JGC C&C

Fluid catalytic cracking is most profitable secondary conversion process in terms of maximizing petrochemical feedstocks as well as fuels. FCC catalyst has a major role in conversion and products yield. Refiners always look for better catalysts and additives as per the operational needs. On an average in every two years, refiners try to change the catalyst to take the advantage of the new formulations as per the changing plant operational philosophy. Evaluation and selection of better catalyst is a continuous activity for researchers to formulate desired catalyst.

CAN FCC™ catalyst technology targets potential deployment at fluid catalytic cracking (FCC) units in Saudi Aramco refineries and affiliates. It is an olefin selective FCC catalyst technology that has jointly developed by Aramco R&DC and JGC C&C. The catalyst is based on innovative Titania- Zirconia inserted USY zeolite as the key active component. This provides higher cracking activity, hydrothermal stability, gasoline, and olefins selectivity. CAN FCC 16 and CAN FCC 17 were developed aiming to maximizing olefin yields in LPG such as propylene and butene.

Fresh catalysts and additive were deactivated and tested using the ACE unit at 530 ◦ C and 4 levels of catalyst to oil ratios 4, 6, 7, and 8. In comparison to the commercial catalyst, the results have shown that CAN FCC 17 demonstrated the highest conversion with approximate 1 wt.%, CAN FCC16 illustrated 2wt.% propylene yield improvement, and all catalysts showed comparable coke yields.

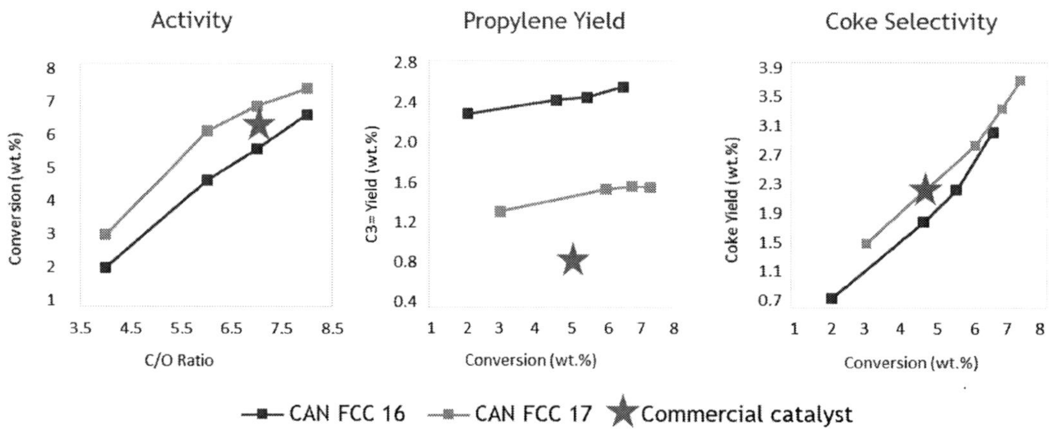

CAN FCC™ Catalysts Performance

Oral 14 (Keynote)
Rational Design of Heterogeneous Catalyst for the Production of Sustainable Chemicals

Javier Martinez
Physical Science and Engineering Division (PSE)
King Abdullah University of Science and Technology
Thuwal 23955-6900, Saudi Arabia
Email: javier.ruizmartinez@kaust.edu.sa

The current production of fuels and chemicals based on fossil feedstocks transformation in a linear fashion is undeniably unsustainable and is bringing concerns on greenhouse gases associated to climate change. Therefore, the conception of new chemical processes based on non-fossil carbon sources will pave the way to a more circular transformation of our economy. Thereby, feedstocks that were considered low-value sources such as biomass, municipal waste and CO_2 will be explored as renewable resources."

The large dissimilarities between fossil and renewable sources call for the design of new chemical and catalytic processes. More specifically, renewable sources are more oxidized and contain a larger number of heteroatoms compared to fossil feedstocks, which are mainly containing hydrogen and carbon. In this contribution, I will show how advanced characterization of catalyst materials by spectroscopy can guide the design of new catalyst materials for the conversion of sustainable feedstock to valuable chemicals.

Oral 15
Synergistic Effects of S-C_3N_4 on Cobalt ferrite Catalysts for Oxygen Evolution Reaction and Enhanced Electrochemical Conversion of Biomass Derived 5-Hydroxymethylfurfural into 2,5-Furandicarboxylic Acid

Maryum Ali[1,2] Muthumariappan Akilarasan[3,4] Muhammad Nawaz Tahir*[1,2] Wasif Farooq*[3,4]

[1]Interdisciplinary Research Center for Hydrogen Technologies and Carbon Management (IRC-HTCM), King Fahd University of Petroleum & Minerals, Dhahran 31261, Kingdom of Saudi Arabia.

[2]Chemistry Department, King Fahd University of Petroleum & Minerals, Dhahran 31261, Kingdom of Saudi Arabia.

[3]Interdisciplinary Research Centre for Refining and Advanced Chemicals (IRC-RAC), King Fahd University of Petroleum & Minerals, Dhahran 31261, Kingdom of Saudi Arabia.

[4]Department of Chemical Engineering, King Fahd University of Petroleum & Minerals, Dhahran 31261, Kingdom of Saudi Arabia.

**Email: muhammad.tahir@kfupm.edu.sa; wasif@kfupm.edu.sa*

The development of efficient and stable electrocatalysts for the oxidation of biomass-derived molecules is critical for advancing sustainable energy technologies. In this study, CoFe and CoFe/S-C_3N_4 composites were synthesized via a hydrothermal method and systematically characterized for their structural and morphological properties using XRD, SEM, and TEM. The materials were employed as electrocatalysts for the oxidation of 5-hydroxymethylfurfural (HMF) to 2,5-furandicarboxylic acid (FDCA) in an H-type electrolyzer equipped with a Nafion membrane separator, using 1 M KOH (pH 13.84) as the electrolyte. Electrochemical studies revealed that the CoFe/S-C_3N_4 composite achieved an overpotential of 100 mV at 10 mA cm^{-2} in the presence of HMF, outperforming CoFe, which required 130 mV. Tafel slope analysis demonstrated enhanced reaction kinetics for CoFe/S-C_3N_4, with lower slopes for both oxygen evolution reaction (OER) and HMF oxidation reaction (HMFOR). HPLC analysis confirmed a high HMF conversion of 98.24%, with a FDCA yield of 90.92% and a Faradaic efficiency of 87.82%. The CoFe/S-C_3N_4 electrode also exhibited excellent long-term and cyclic stability, retaining 96.26% HMF conversion after three cycles. These findings highlight the potential of CoFe/S-C_3N_4 as a highly efficient and durable catalyst for biomass valorization and renewable energy applications.

Oral 16
Tuning the Properties of ZSM-5 Additive for Co-cracking of Waste Plastics Dissolved in Vacuum Gas Oil

Abdulkadir Tanimu*, Abdullah Aitani, Ramzi Hadi Al-Shuqaih

Interdisciplinary Research Center for Refining and Advanced Chemicals, Research Institute, King Fahd University of Petroleum and Minerals, Dhahran 31261, Saudi Arabia.

Mesoporous ZSM-5 zeolites were synthesized by hydrothermal rapid ageing approach at high stirring temperature of 40 °C, and the Si/Al ratio of the zeolite was varied. Thus, 15 wt.% ZSM-5 zeolites of various textural properties, morphologies, crystal sizes and acidity were mixed with equilibrium catalyst (E-Cat) to form the catalyst formulations. The activity of these mesoporous ZSM-5 zeolites and commercial ZSM-5 additive (ZSM-5(COM)) was studied in the co-cracking of low-density polyethylene (LDPE) and heavy vacuum gas oil (HVGO) feedstock in a microactivity test (MAT) unit under different conditions of temperature and loadings of LDPE. The results showed that conversion and light olefins yield increased with the increase in the loading of LDPE from 2.5 – 10 wt.% over E-Cat. The cracking of LDPE/HVGO over E-Cat/ZSM-5(COM) showed slight decrease in the conversion for the HVGO and 2.5LDPE/HVGO, and this is due to the small pore size of ZSM-5(COM) that hindered the diffusion of heavy molecules of HVGO to the active sites. However, in 10LDPE/HVGO, the conversion increased significantly. The yield of light olefins increased significantly for all the feeds, and this is ascribed to the shape selectivity of ZSM-5(COM). Compared to the E-Cat/ZSM-5(COM) with conversion of 75%, the E-Cat/ZSM-5(50) showed increased conversion of 83.4% in the cracking of 2.5LDPE/HVGO, with more naphtha and light olefins. This is ascribed to the mesoporous structure of ZSM-5(50) that enhanced diffusion of heavy molecules of HVGO to the active sites of the zeolite and its balanced acid sites that prevent excessive cracking and hydrogen transfer reaction. The cracking of 2.5LDPE/HVGO over E-Cat/ZSM-5(25) resulted in conversion of 75.5%, which increased to 81% in the case of E-Cat/ZSM-5(75), and the total gas yield increases from 39.2% to 51.1%, where the light olefins increased from 29.8% to 38.4%. The ethylene and propylene yields increased from 3.6% and 14.7% to 4.8% and 19.9% respectively. The ZSM-5(75) has more strong acid sites (0.107 mmol/g) that cracked further the naphtha range to light olefins, and its optimum pore dimension hindered diffusion of the naphtha range product out of the zeolite internal cavity, thus exposing them to the strong acid sites for selective cracking to light olefins.

Oral 17
Catalytic Microwave Assisted Pyrolysis of Waste Plastics

Abdul Gani Abdul Jameel
Department of Chemical Engineering, KFUPM; Interdisciplinary Research Center for Refining and Advanced Chemicals, KFUPM

Microwave-assisted catalytic pyrolysis presents a promising approach for converting petroleum-based plastic waste into valuable chemicals, advancing the goals of a sustainable circular economy. This study investigates the catalytic pyrolysis of low-density polyethylene (LDPE) under microwave irradiation, focusing on the impact of various catalysts (HZSM-5, Ga/ZSM-5, Ga/Ni/ZSM-5, Ga/Co/ZSM-5, and Ga/Cu/ZSM-5) on product yield and distribution. Among these, Ga/ZSM-5 achieved the highest liquid oil yield (41%), while Ga/Ni/ZSM-5 excelled in producing long-chain olefins, contributing 27% of the liquid fraction. In contrast, Ga/Co/ZSM-5 predominantly generated heavy pyrolysis oil with 25% long-chain paraffins, making it less suitable for high-value chemical production. Notably, Ga/Cu/ZSM-5 produced an aromatic-rich pyrolysis oil, with benzene derivatives constituting 90% of the liquid fraction, demonstrating its effectiveness for targeted applications. SimDist analysis compared the liquid product distribution with a petroleum assay, highlighting that HZSM-5 could produce an 86.4% naphtha fraction. Additionally, Ga/Cu/ZSM-5 catalyzed the highest production of hydrogen and syngas, as confirmed by MicroGC analysis, while also exhibiting the highest coke deposition (1.35%) due to its elevated aromatic hydrocarbon content and hydrogen release. Fresh and spent catalyst properties were analyzed to understand activity retention and the effects of metal doping on product distribution. These results underscore the potential of microwave-assisted catalytic pyrolysis, particularly with Ga/Cu/ZSM-5, for efficiently transforming plastic waste into valuable chemicals, contributing to resource sustainability and environmental conservation.

Oral 18 (keynote)
Hydrogenation of CO_2 for Carbon Neutrality: Enhancing Efficiency with Membrane Reactors

Masahiko Matsukata
Department of Chemical Engineering, Waseda University
513 Wasedatsrumaki-cho, Shinjuku-ku, Tokyo 162-0041, Japan
Tel & Fax: +81-3-5286-3850, E-mail: mmatsu@waseda.jp

The escalating global climate crisis, primarily driven by anthropogenic greenhouse gas emissions, necessitates urgent and effective measures to mitigate climate change. Carbon neutrality has emerged as a pivotal goal to achieve a sustainable future. The hydrogenation of CO_2 offers a promising pathway to convert this greenhouse gas into valuable chemicals and fuels, thereby contributing to carbon capture and utilization (CCU) strategies.

Various hydrogenation reactions, including the reverse water-gas shift reaction, methanation, methanol synthesis, and Fischer-Tropsch synthesis, have been extensively studied. However, these processes are often hindered by thermodynamic limitations, catalyst deactivation, and product selectivity issues. A common challenge in these reactions is the formation of water, which can lead to catalyst oxidation, sintering, and equilibrium limitations.

To overcome these challenges and enhance the efficiency of CO_2 hydrogenation, in situ water removal is essential. Membrane reactors, which integrate catalytic reaction and membrane separation, offer a promising solution. By continuously removing water from the reactor, membrane reactors can shift the reaction equilibrium, suppress side reactions, and prolong catalyst lifetime. Moreover, membrane reactors can facilitate the production of high-purity products.

This presentation will propose the fundamental principles of membrane reactors and their application to CO_2 hydrogenation. The advantages of membrane reactors over conventional fixed-bed reactors will be discussed, including improved reaction rates and stability. Furthermore, the presentation will explore the potential of membrane reactors in addressing grand challenges such as carbon neutrality and energy sustainability.

Oral 19
Steering electrochemical CO_2 reduction toward liquid fuels and chemicals

Husain Naji
Saudi Aramco

Given the rapidly rising concentration of carbon dioxide in the atmosphere and its influence on global temperature, the need for a sustainable pathway toward CO_2 capture and reduction is increasingly becoming more urgent. Electrochemical CO_2 reduction has emerged as an attractive approach to utilize the waste CO_2 gas as a valuable feedstock to produce valuable fuels and chemicals.

Pulse electrolysis has emerged recently as simple and effective technique to control product selectivity and enhance catalyst stability. This technique allowed researchers to improve CO_2 mass transport, modify catalyst nanostructure, oxidation state, and facets, and control interfacial pH and adsorbate coverage, allowing for unprecedentedly high product selectivity toward hydrocarbon products.

Here, we demonstrate the use of pulse electrolysis to control catalyst morphology and oxidation state to produce liquid fuels with high activity and selectivity. An aqueous flow-through electrochemical cell is developed to reduce CO_2 at commercial current densities with high stability toward multi-carbon products. Various operating parameters are investigated such pulse duration, applied potentials, and metal doping are investigated to optimize the production of valuable liquid products. Overall, this work is a step forward in the production of E-fuels directly from the captured CO_2, fostering the transition toward the circular carbon economy.

Oral 20

Development of Highly Dispersed Pd Nanoparticles Supported on Dendritic Mesoporous Heterogeneous Catalysts for Efficient CO_2 Hydrogenation to Methanol

Mohnnad H. Alabsi[1]*, Lujain K Bashihab [1]
[1] *Advanced Materials & Catalysis, Aramco Research Center, Thuwal, 23955-6900 Saudi Arabia.*
**Email: Mohnnad.alabsi@aramco.com*

A Pd2.0/CZZ catalyst (CeZrZnOx supported Pd catalyst with 2.0 wt% Pd) with dendritic morphology and highly dispersed Pd was prepared for methanol (MeOH) synthesis from CO_2 hydrogenation. The deposition–precipitation method, higher specific surface area, and pore volume of CZZ support improve the dispersion degree of Pd on the CZZ support. The ultra-dispersed and optimized amount of Pd plays a crucial role in H_2 dissociation, generating more spillover hydrogen on the Pd active sites, thus improving the CO_2 hydrogenation performance. Moreover, introducing the optimized Pd of the Pd2.0/CZZ catalyst increases Ce^{3+}/Ce^{4+} ratios, which are critical in generating more oxygen vacancies of the CZZ support, thus promoting the adsorption and activation of CO_2 molecules. Pd2.0/CZZ catalyst presents the highest CO_2 conversions (29.1 %), superior MeOH selectivity (43.8 %), highest MeOH yield (12.7 %), highest turnover frequency (TOF) (87.3 h^{-1}), highest STY (5.11 molMeOH $kg^{-1}h^{-1}$) and excellent 100 h long-term stability among a series of Pdx/CZZ catalysts and the reference PdZn/CZ-Co catalyst. The in-situ DRIFTS and density functional theory (DFT) simulation suggests that Pd2.0/CZZ catalyst with adequate Pd particle size and relatively high Pd metal dispersion promotes the pathway to *HCOO and $*CH_3O$ as primary intermediates of CO_2 hydrogenation to MeOH.

Poster 01

Post-synthesis functionalization of covalent organic framework for CO_2 capture from DAC

Mona Al-Otaibi[1], Mahmoud Abdelnaby[2], Mohamed Essalhi[2], Ammar AlAhmed[1]
[1] Carbon management Research Division, Research & Innovation Complex, Saudi Aramco, Dhahran 5022, Saudi Arabia
[2] Interdisciplinary Research Center for Hydrogen Technologies and Carbon Management (IRC-HTMC), King Fahd University of Petroleum & Minerals, Dhahran 31261, Saudi Arabia

Carbon dioxide direct air capture (DAC) is a crucial part of the attempts to slow global warming. DAC requires a highly selective system to sort out an extremely low number of CO_2 molecules (~400 ppm) among other dominant components (N_2, O_2) in the air. The need for a substance to have a reactive CO_2-specific sites is necessary to extract CO_2 from such a diluted source. Aqueous solutions of mono-ethanolamine have already been used for this separation. However, using them results in high energy consumption and has operational issues including decomposition and toxicity.

In order to lower the energy consumption compared to aqueous amine solutions, many classes of solid-state materials have been explored for this purpose. ARAMCO in collaboration with KFUPM work on developing porous-nature material termed covalent organic frameworks (COF's) and their functionalization with aliphatic amines. Here, we present the utilization of (COFs) as solid-state platforms for the covalent bonding of amines that can be employed to capture CO_2 from the air. Utilizing COFs has advantages for finely tuning framework design and modifying the chemical environment within of their pores. Currently, olefin and dioxane based COFs namely V-COF-1 and COF-316 have been synthesized and modified with amine species to improve their performance towards CO_2 gas. the chemistry of framework conversion and corroborated the accelerated CO_2 uptake phenomena was confirmed through the Fourier transform infrared spectroscopy and solid state nuclear magnetic resonance spectroscopy. Interestingly, these aminated platforms showed an enhance of CO_2 uptake for 40% comparing to the original structure at 400 ppm. The developed functionalized COFs showed exceptional stability in the CO_2 adsorption desorption multicycle measurements which make them promising candidates as solid sorbent for DAC.

Poster 02

DFT study of CO_2 Hydrogenation Over Alumina-Based Catalysts via RWGS Reaction

Abdulrahman H. Yassin, Mohnnad H. Alabsi, and Mohammed A. Daous
KAUST R&D, Saudi Aramco, Thuwal 23955, Saudi Arabia

The conversion of CO_2 gas into e-fuel has emerged as a prominent focus within the energy sector, offering a potential pathway to realizing the goal of net-zero emissions. For example, hydrogen and CO_2 undergo transformation via the reverse water-gas shift (RWGS) reaction to produce syngas. This study seeks to innovate a novel catalyst specifically tailored for the RWGS process. Alumina-based materials have demonstrated encouraging outcomes in existing research, thereby positioning them as highly promising candidates for the development of our new RWGS catalyst. Here, the aim is to investigate computationally the mechanisms of CO_2 conversion on γ-alumina and γ-alumina surfaces to have a deep understanding of this chemical reaction. CP2K program has been utilized to perform the density functional theory (DFT) simulations of the reaction of CO_2 molecules with slabs of Al_2O_3 using the PBE functional with D3-BJ dispersion correction. The adsorption of CO_2 molecules on ?-alumina and ?-alumina surfaces has been performed as well as the effect of K as a promoter. The results clarified how we can tune the chemical reaction to go in the RWGS pathway to produce more syngas and avoid bond cleavage or formate formation pathways, which consequently enhance the production of unwanted methane gas. The effect of additional metals will be performed in the future studies to enhance the conversion of CO_2 molecules into syngas.

Poster 03

Enhancing Hydrogen Production: Acidic-Basic Structural Modification of Nickel-based catalysts for Ammonia Decomposition

Reem Albashrawi [1*], Emad N. Al-Shafei [1], Mohammad Z. Albahar [1], Mohammad Aljishi [1], Ahmed Alasseel [1] , Gazali Tanimu[2], Abdullah Aitani [2]

A[1] Research and Development Center, Saudi Aramco, Dhahran, Saudi Arabia
[2] Center for Refining & Advanced Chemicals, King Fahd University of Petroleum & Minerals, Dhahran 31261, Saudi Arabia
**Email: Reem.Albashrawi@gmail.com*

The increase in global sustainable green energy and hydrogen fuel trend towards net zero emission for the transport sector expected to surge hydrogen production. The decarbonization of energy processes will boost the use of hydrogen carriers like ammonia, methanol and methyl-cyclo-hexane. Ammonia seems to be an excellent choice, due to its high hydrogen content compared to other hydrocarbon streams. However, the catalytic decomposition of ammonia is still challenging in term of sustainable catalyst development. Ammonia decomposition requires several steps comprising adsorption over active catalyst/support, enough binding energy for nitrogen-hydrogen (N-H) bond activation for hydrogen abstraction via dehydrogenation followed by desorption of N_2 and H_2. Several catalysts were explored for ammonia cracking showed the most active catalysts. However, these catalysts required selected supports and promoters to upsurge the catalytic activity for greater ammonia conversion and lowering the cracking temperature. The objective of the study was to modify the acid-base surface of catalyst structure, by impregnating metal promotors of Na, K and Ba to Ni oxide and using moderate acidic surface support of gamma alumina for ammonia conversion [1]. Metallic oxides of Co and La were also used to enhance surface basicity by introducing transition and rare earth metals to incorporate with Ni oxide for better dehydrogenation of ammonia at lower temperature. The modification of Ni oxide structure with promoter is expected to enhance bifunctionality of the catalyst by expanding promotor-metal-support interaction. The catalytic reactivity over trimetallic catalysts was systematically influenced by their basic-acidic property that promoted ammonia dehydrogenation and catalyst activity.

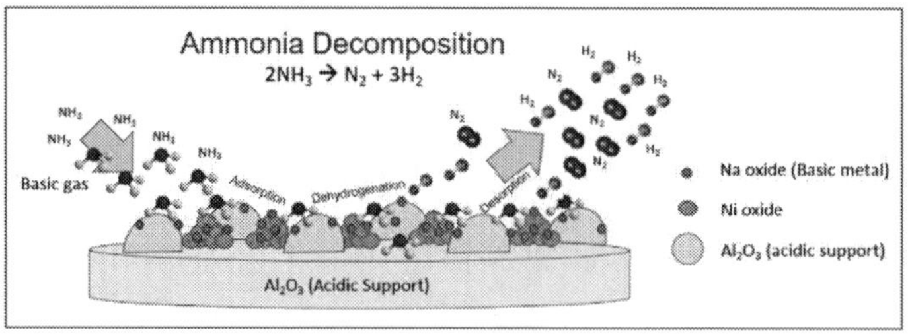

Reference:
1] Al-Shafei, E.N., Albahar, M.Z., Albashrayi, R., Aljishi, M., Alasseel, A., Tanimu, G. and Aitani, A., 2023. The effect of acidic–basic structural modification of nickel-based catalyst for ammonia decomposition for hydrogen generation. *Molecular Catalysis*, *550*, p.113581.

Poster 04

Electrochemical Ammonia Decomposition over Nickel and Cobalt Nanostructures on Nickel Foam for Sustainable Hydrogen Production

Umar Jafar [a], Mustapha Umar [b], Yahaya Gambo [c], Khalid Alhooshani [c,d], and Yousef Alsunni [a,c]

[a] *Department of Chemical Engineering, King Fahd University of Petroleum and Minerals, Dhahran, 31261, Saudi Arabia.*

[b] *Department of Chemical Sciences, North-Eastern University, Gombe, 086, Nigeria.*

[c] *Interdisciplinary Research Center for Refining and Advanced Chemicals, King Fahd University of Petroleum and Minerals, Dhahran, 31261, Saudi Arabia.*

[d] *Department of Chemistry, King Fahd University of Petroleum and Minerals, Dhahran, 31261, Saudi Arabia.*

Hydrogen, recognized as a clean energy alternative, encounters significant challenges related to storage and transportation [1]. Ammonia emerges as a promising hydrogen carrier that addresses these issues. Its efficient electrochemical decomposition (EC) into hydrogen demands less energy compared to other methods, positioning it as a viable pathway for a sustainable energy future [2]. Herein, we investigated the EC of ammonia for hydrogen production using nickel foam (NF) as a catalytic substrate. Nickel and cobalt nanostructures were synthesized on NF via a hydrothermal method. Comprehensive characterization, including SEM, AFM, XRD, Raman, FTIR, and XPS, revealed significant morphological and chemical modifications of the NF surface, suggesting enhanced catalytic activity for ammonia decomposition. The electrochemical performance of the modified NF electrodes was evaluated through LSV, CV, EIS, controlled potential coulometry, and chronopotentiometry. The ammonia oxidation reaction (AOR) activity of these electrodes was investigated in an alkaline electrolysis cell using a solution of 0.5 M NH_4OH and 1.0 M KOH at ambient conditions. Both Ni/NF and Co/NF exhibited significantly enhanced catalytic activity compared to bare NF, evidenced by lower onset potentials, higher current densities, and reduced charge transfer resistance [1-4]. Notably, Co/NF demonstrated superior performance, achieving higher ammonia conversion, hydrogen production rates, and energy efficiency. Additionally, Density Functional Theory (DFT) calculations were performed to elucidate the underlying reaction mechanisms. Utilizing a modified computational hydrogen electrode (CHE) approach, we determined an energy span of 2.27 eV for Co/NF, which is lower than the 2.45 eV observed for Ni/NF. Moreover, it was found that the ammonia adsorption is more favorable on Ni surfaces, while nitrogen desorption is more favorable on Co surfaces, elucidating the roles of each electrode in facilitating AOR. Overall, these results highlight the potential of nanostructured metal modifications to significantly improve the catalytic activity and overall performance of NF electrodes for efficient ammonia oxidation, offering a promising pathway for sustainable hydrogen production.

Poster 05

Maximizing CO Production from CO_2 Conversion for E-Fuel: Reverse Water-Gas Shift Reaction

Emad Al-Shafei [1*], Mohammad Alalouni [1], Gazali Tanimu [3,4], Reem Albashrawi [1], Eman Albahar [1], Mohammad Aljishi [1], Ahmed Alasseel [2], Abdullah Aitani [3,4]

[1]*Catalyst Center of Excellence, Research and Development Center, Saudi Aramco, Dhahran 31311, Saudi Arabia.*

[2] *Research and Analytical Services Department, Saudi Aramco, Dhahran 31311, Saudi Arabia.*

[3] *Center for Refining & Advanced Chemicals, King Fahd University of Petroleum & Minerals, Dhahran 31261, Saudi Arabia*

[4] *Chemical Engineering Department, King Fahd University of Petroleum & Minerals, Dhahran 31261, Saudi Arabia*

**Email: emadnaji@gmail.com*

The development of catalysts for the Reverse Water-Gas Shift (RWGS) reaction is essential due its significant potential in addressing climate and sustainability goals through the conversion of CO_2, a significant greenhouse gas, into CO, which can be further processed into valuable chemicals and fuels. Although RWGS reaction can operate at a range of temperatures to produce CO with varying selectivities, it tends to favor CO_2 methanation pathway at lower temperatures, resulting in methane formation. This study investigates the performance of Rh/ TiO_2 and Cu/ TiO_2 catalysts, which are known for their ability to dissociate H_2 and activate CO_2, for the application of RWGS under a range of temperatures (500°C <T< 700°C). At a CO_2/H_2 ratio of 1:3, Cu/ TiO_2 catalyst achieved 100% CO selectivity, while Rh/ TiO_2 catalyst produced 82% CO alongside 18% methane. When the CO_2/H_2 ratio switched to 3:1, RWSG reaction shifted to highest selectivity of CO (99.5%) over Rh/ TiO_2 resulting in minimal methane formation. Kinetic studies revealed first-order behavior for Rh/ TiO_2 and second-order behavior for Cu/ TiO_2, indicating different reaction mechanisms of RWGS over each catalyst. The results highlighted the importance of optimizing the CO_2/H_2 ratio and reaction temperature to maximize CO yield and minimize undesired byproducts. This research provides important insights into improving industrial catalysts for efficient CO_2 utilization.

Poster 06

Quantitative Risk Assessment of Hydrogen Refueling Stations: A Case Study for Saudi Arabia

Bashir Hashim[1], Sunhwa Park[1,2 *]
1 Chemical Engineering Department, King Fahd University of Petroleum Minerals, Dhahran, Eastern Province, Saudi Arabia 2 KFUPM Interdisciplinary Research Center for Refining and Advanced Chemical, Dhahran 31261, Saudi Arabia
**Email: sunhwa.park@kfupm.edu.sa*

The global expansion of hydrogen refueling stations (HRSs) is essential for advancing the adoption of fuel cell electric vehicles (FCEVs) and transitioning to a sustainable energy future. This study presents a comprehensive framework for evaluating HRS safety under the unique environmental and operational conditions of Saudi Arabia. By integrating consequence-based and risk-based analyses, the research quantified safe distances for key incident scenarios, including jet fires and vapor cloud explosions (VCEs), using GEXCON EFFECTS and RISKCURVE software. The results highlight significant variability in harm zones across six critical nodes, with compressors and dispensers posing the highest risks. Individual risk (IR) contours revealed that compressors, intermediate cylinders, and dispensers contributed over 80% of the total risk. Societal risk (SR) using Netherland's ALARP criteria was analyzed, and the total risks fall within the ALARP region necessitating the need for mitigating measures. These findings emphasize the need for targeted mitigation strategies, including advanced leak detection, robust isolation mechanisms, and strategic node placement. This study offers actionable insights for policymakers and stakeholders, advancing HRS design, operation, and public trust in hydrogen infrastructure.

Poster 07

Selecting Hydrogen Storage in Saudi Arabia: A Multi-Criteria Framework Integrating FAHP for Uncertainty and Safety

YooJeong Oh[1], Hans Pasman[2], Safyan Akram Khan[4], Sunhwa Park[1,3*]
[1] *Chemical Engineering Department, King Fahd University of Petroleum and Minerals (KFUPM), Dhahran 31261, Saudi Arabia*
[2] *Mary Kay O'Connor Process Safety Center, Texas A&M Engineering Experiment Station, College Station, TX, 77843-3122, USA*
[3] *KFUPM Interdisciplinary Research Center for Refining and Advanced Chemical, Dhahran 31261, Saudi Arabia*
[4] *KFUPM Interdisciplinary Research Center for Hydrogen Technologies and Carbon Management, Dhahran 31261, Saudi Arabia*

This study introduces a comprehensive multi-criteria decision-making (MCDM) framework tailored to assess hydrogen storage technologies in the context of Saudi Arabia's unique geographical, industrial, and climatic factors. The research integrates Interval-Valued Pythagorean Fuzzy Sets (IVPFS) with two decision-making techniques—the Analytic Hierarchy Process (AHP) and the Technique for Order Preference by Similarity to Ideal Solution (TOPSIS). Through this integrated approach, the study evaluates five key hydrogen storage options: (1) compressed hydrogen gas, (2) liquefied hydrogen, (3) cryo-compressed hydrogen, (4) underground hydrogen storage, and (5) chemical storage. Thirteen criteria are assessed across four dimensions—technological, economic, social, and safety—ensuring a holistic analysis that aligns with stakeholder priorities. The findings indicate that underground storage is the most viable large-scale solution for Saudi Arabia, thanks to its high storage capacity, lower operational costs, and compatibility with geological formations such as salt caverns and depleted reservoirs. Compressed and liquefied hydrogen storage options also show promise but faces challenges related to Saudi Arabia's hot climate, which necessitates high pressure and cooling. Although chemical storage offers enhanced safety, its adoption is hindered by high costs. This framework provides decision-makers with a systematic approach to evaluate hydrogen storage technologies, balancing diverse considerations essential for Saudi Arabia's strategic shift toward a sustainable, hydrogen-focused energy future.

Poster 08

Exploring the Synergy Between CoP and Co_2P Phases: A Comparative Study of Cobalt Phosphides phases and Their Composite for Hydrogen Evolution Reaction

Wajiha Fatima [1], Tarek. A. Kandiel [1,2]
[1] *Department of Chemistry, King Fahd University of Petroleum and Minerals (KFUPM), Dhahran,31261, Saudi Arabia*
[2] *Interdisciplinary Research Center for Hydrogen Technologies and Carbon Management (IRC-HTCM) at KFUPM, Dhahran, 31261, Saudi Arabia*
**Email: g202216100@kfupm.edu.sa, wajiha9292@gmail.com*

Transition metal phosphides have emerged as viable, earth-abundant alternatives to platinum for catalyzing the hydrogen evolution reaction (HER) in acidic electrolytes. This study investigates cobalt phosphide nanoparticles as electrocatalysts for the hydrogen evolution reaction (HER). Cobalt phosphides are available in many compositions, and the synthesis of unique phases is highly significant due to their wide-ranging applications in electrocatalytic reactions. The phase-selective synthesis of cobalt phosphides, comprising pure Co2P, CoP, and a CoP/ Co2P composite, was accomplished through the thermal decomposition of cobalt acetate tetrahydrate $[(CH_3COO)_2{\cdot}4H_2O]$, utilizing trioctylphosphine as a phosphorus precursor, with oleylamine and 1-octadecene serving as solvents and stabilizing agents. Controlled phase formation was effectively achieved by systematically adjusting the ratio of trioctylphosphine to cobalt acetate, together with the temperature and reaction duration.The synthesized phases of cobalt phosphides, Co2P, CoP, and the CoP/ Co2P composite, were assessed as electrocatalysts for hydrogen evolution reaction (HER) in a 0.5 M H_2SO_4 solution to establish a structure-to-activity link.Our results indicated that Co2P, CoP, and the CoP/ Co2P composite are all suitable candidates for hydrogen evolution reaction (HER), with the composite exhibiting superior performance owing to its optimized phase interaction. This study demonstrates that the catalytic efficacy of transition metal phosphides can be markedly enhanced by optimizing active phases, emphasizing their use in watersplitting technologies

Poster 09

Decorating $Zr_{12}O_{12}$ nanocage with transition metals (Ti, V, Cr, or Mn) to act as a single-atom catalyst for water splitting application

Sajjad Hussain[a,b], Abdulaziz A. Al-Saadi[a,b,*]
[a] Department of Chemistry, King Fahd University of Petroleum & Minerals (KFUPM), Dhahran 31261, Saudi Arabia
[b] Interdisciplinary Research Center for Refining and Advanced Chemicals, King Fahd University of Petroleum & Minerals, Dhahran 31261, Saudi Arabia

Catalytic water splitting represents a highly promising approach to address the increasing global demand for sustainable and clean hydrogen energy. Various metal oxide semiconductors, such as ZnO, TiO_2, Fe_2O_3, MnO, $BiVO_4$, and ZrO_2, have been developed to lower the activation energy and efficiently catalyze water splitting reactions leading to hydrogen and oxygen. ZrO_2 stands out as a highly promising photocatalyst, owing to its exceptional electronic band structure and remarkable physicochemical stability. However, its wide band gap and rapid recombination of charge carriers limit its catalytic activity. Nanocages with hollow interiors and ultrathin walls could be an effective strategy to promote the performance of ZrO_2 catalysts due to their favorable specific surface area and efficient channels for mass transport and reactant molecular contact. Herein, the role of $Zr_{12}O_{12}$ and M@$Zr_{12}O_{12}$ (M = Ti, V, Cr, or Mn) has been investigated in the overall water splitting process using first-principles calculations. Density functional theory (DFT) simulations reveal that the M@$Zr_{12}O_{12}$ cages demonstrates a high stability, which is crucial in designing single atom catalysts (SACs) to prevent aggregation of the metal. From electronic structure calculations, it has been observed that the loading of metal atoms reduces the bandgap of the $Zr_{12}O_{12}$ nanocage, thereby increasing the visible light absorption of the catalysts. Additionally, the computed Gibbs free energies indicate that the Mn@$Zr_{12}O_{12}$ nanocage exhibits a superior ability in hydrogen evolution reaction (ΔGH^* = -0.34 eV) at room temperature among the M@$Zr_{12}O_{12}$ catalysts explored.

Poster 10

Production of Hydrogen and Value-Added Chemicals through Microwave Assisted Pyrolysis (MAP) of Arabian Crude Oil

Intisar Ul Hassan[1], Aniz Ummer[2], Abdul Gani Abdul Jameel[1,2], *
1Department of Chemical Engineering, King Fahd University of Petroleum and Minerals, Dhahran 31261, Saudi Arabia 2 Interdisciplinary Research Center for Refining and Advanced Chemicals, King Fahd University of Petroleum and Minerals, Dhahran 31261, Saudi Arabia
**Email: a.abduljameel@kfupm.edu.sa*

Hydrogen (H_2) production using traditional methods, such as steam methane reforming (SMR), relies on fossil fuels and operates at high temperatures, with hydrocarbons as the main feedstock. However, these processes release large amounts of carbon dioxide (CO2) as a byproduct, requiring costly carbon capture and storage systems to mitigate environmental impact. This study explores an alternative approach to hydrogen production through the catalytic pyrolysis of Arabian crude oils at moderate temperatures. Using microwave heating, this study facilitates reactions between crude oil and catalysts in a batch reactor system. The research focuses on three crude oil types-Arab Heavy (API gravity 27°), Arab Light (API gravity 34°), and Arab Extra Light (API gravity 39°) at temperatures ranging from 400 to 500 °C. Ni based catalysts will be used, and the effects of varying catalyst-to-oil (C/O) ratios will be studied. This method is expected to increase hydrogen production and improve the yields of valuable byproducts such as light olefins (C2–C4), alkanes, and aromatic compounds. To analyse the results, gas composition will be measured using Micro-GC, and the quality of the pyrolysis oil will be assessed using techniques like Simulated Distillation (SimDist), GC-MS, 1HNMR, and FTIR spectroscopy. Additionally, the performance of fresh and spent catalysts will be evaluated through BET surface area analysis, X-ray diffraction (XRD), thermogravimetric analysis (TGA), scanning electron microscopy (SEM) for surface morphology, and EDS for elemental composition. This work aims to demonstrate the feasibility of producing hydrogen and value-added chemicals directly from crude oils, offering a potentially more sustainable and economically viable alternative to traditional hydrogen production methods.

Poster 11

Dry Reforming of Methane to Syngas using Ni/Al_2O_3 catalysts Promoted by Lanthanides

Imtiaz UL Hasan[1*], Zuhair Malaibari[1,2], Shakeel Ahmed[2]
[1] *Department of Chemical Engineering, King Fahd University of Petroleum and Minerals, Dhahran 31261, Saudi Arabia*
[2] *Interdisciplinary Research Center for Refining and Advanced chemicals, King Fahd University of Petroleum and Minerals, Dhahran 31261, Saudi Arabia*

**Email: g202315990@kfupm.edu.sa*

Dry reforming of methane (DRM) is the process to produce syngas using greenhouse gases, methane and carbon dioxide in the presence of metal catalyst. Syngas, which is a mixture of CO and H2, is produced from various resources such as biomass, naphtha, coal and natural gas. Steam methane reforming (SMR), partial oxidation of methane (POX) and autothermal reforming technologies are used to produce syngas with H2/CO ratio, 3, 2, and 1.6-1, respectively. However, DRM produces syngas with H2/CO = 1, that is used in Fischer Tropsch synthesis to produce other liquid hydrocarbons [1]. The major problem associated with DRM is the formation of coke, on catalyst surface. Methane decomposition and Boudouard reaction produces carbon which sticks on the catalyst surface and deactivates the active metal. Carbon is produced in the form of pyrolytic, whisker and gum, pyrolytic & whiskers are favored at high temperature while gum is at lower temperature. The high temperature and acidity of the catalytic sites are the main reasons behind the carbon formation, which can be decreased by using the noble metals such as Pt, Pd, and Rh with the Al2O3 or SiO2, but noble metals are expensive. Ni serves as a cost-effective alternative to noble metals like Pt, Pd, and Rh for catalyzing DRM with minimal loss in performance. However, Ni catalysts tend to deactivate quickly under practical DRM conditions. Rare earth oxides, such as CeO2 or CeO2–ZrO2–Al2O3 (CZA), are often employed as supports to enhance the activity and stability of Ni catalysts due to their redox properties. Despite these advantages, redox-active supports can promote the reverse water gas shift (RWGS) reaction, which reduces hydrogen selectivity [2]

Poster 12

Catalytic Performance in Dry Reforming of Methane: A Comparative Study of SiO_2 Encapsulated and Unencapsulated Ni/Cao-Al_2O_3 Catalysts

Sherif Alabi [a], Sagir Adamu [a,b], Abdallah A. Al-Shammari [a,b], Mohammad M. Hossain [a,b]
a. Department of Chemical Engineering, King Fahd University of Petroleum & Minerals, Dhahran, Saudi Arabia
b. Interdisciplinary Research Center for Refining and Advanced Chemicals, King Fahd University of Petroleum & Minerals, Dhahran, Saudi Arabia

**Email: g202303930@kfupm.edu.sa*

Advancement in catalyst design for CO_2 dry reforming has surge for the search for an innovative and sustainable pathway for converting CH_4 and CO_2, into syngas. This enables the conversion of syngas into valuable chemicals and fuels. DRM offers an alternative approach for hydrogen production and other energy products which are in alignment with reduction of emission in the environment. However numerous challenges have been faced on the deactivation, stability and coke resistance of the catalyst. To address these issues, recent advancements have focused on developing robust catalyst structures, notably through silica encapsulation. This approach involves the encapsulation of Ni-based catalysts in a mesoporous silica shell which creates a barrier that inhibits Ni sintering and reduces coke formation.

In this study, we investigated the effect of silica encapsulation on Ni/Cao-Al_2O_3 catalysts. Ni/CaO-Al_2O_3 was prepared via impregnation, and the SiO_2 encapsulation process was carried out by adopting the Stöber method to create a silica shell over the catalyst (denoted as Ni/CaO-Al_2O_3@SiO_2). The catalyst was screened and evaluated for performance at 650 ^{o}C using the CREC riser simulator, focusing on CH_4 and CO_2 conversion, as well as H_2/CO selectivity and yield. XRD, H_2-TPR, Raman Spectroscopy, TGA and SEM techniques are used for catalyst characterization for in-depth understanding of the effect of encapsulation. From the catalyst activity test, Ni/CaO-Al_2O_3@SiO_2 demonstrated significantly higher conversion CH_4, and CO_2 as compared with Ni/CaO-Al_2O_3 this can be attributed to the improved metal-support interaction in the encapsulated catalyst. Furthermore, as demonstrated by from the findings, the Ni/CaO-Al2O_3@SiO_2 catalyst significantly increased coke resistance and decreased catalyst sintering. These findings highlight the critical role of silica encapsulation as an approach of tuning catalytic properties in other to increase resistance to coke formation and higher catalytic activity.

Poster 13

Synthesis of ZSM-5 Nanorods for Low Temperature Carbon Dioxide Reforming of Methane to Syngas

Yahuza N. Abdulai [a], Ijaz Hussain [b], Khalid Alhooshani [a, b], Saheed A. Ganiyu [a, b*]
[a]*Department of Chemistry, King Fahd University of Petroleum and Minerals, Dhahran 31261, Saudi Arabia.*
[b] *Interdisciplinary Research Center for Refining & Advanced Chemicals, King Fahd University of Petroleum & Minerals, Dhahran 31261, Saudi Arabia*
**Email: gsadewale@kfupm.edu.sa*

Syngas is produced by carbon dioxide reforming of methane (DRM), which is a sustainable method of using greenhouse gases and generating useful chemical feedstocks. However, due to coke formation and sintering of nickel-based catalysts, this often leads to rapid deactivation of catalysts at DRM temperatures (700-1000 °C), making the commercialization of DRM technology very difficult. In this work, the DRM performance of a highly stable Ni/ZSM-5 catalyst was evaluated using a tubular stainless-steel fixed bed reactor. The support, ZSM-5 was synthesized by employing the seed assisted hydrothermal method. The Ni/ZSM-5 catalyst was synthesized by innovative wet impregnation techniques. SEM and TEM morphological assessment of the ZSM-5 support revealed well-defined knife-edged nanostructuring. The Ni/ZSM nanorod was further characterized by XRD, N2 adsorption-desorption techniques, H_2-TPR, CO_2-TPD, XRF and solid-state NMR. The optimized catalyst exhibited remarkable conversion with constant rates of methane and carbon dioxide conversion over a broad temperature range (500-1000 °C), and H_2/CO ratio suitable for downstream methanol synthesis. Raman, TGA and TPO characterization of the spent catalysts revealed less coke formation. According to these findings, the Ni/ZSM-5 nanorod is a stable and efficient catalyst for DRM, offering a sustainable method of producing syngas while tackling issues with carbon management with high potential for industrial applications.

Poster 14

Case Study on The Acid Sites Formation and Morphology in Amorphous Silica-Alumina

Batool Altaher, Lianhui Ding, Mohammad Bahhar, Faisal Mulla
Catalyst Center of Excellence, R&DC, Saudi Aramco, Dhahran, Saudi Arabia

Several methods have been deployed to prepare composite materials as catalyst supports for hydrocracking. Amorphous silica alumina is an inorganic material with high surface area and pore volume which makes it a potential candidate with high activity in the hydrocracking applications. As zeolite microporosity can be beneficial in term of shape selectivity, it has inherent diffusion limitations that result in slow mass transport and reduced reaction rates that can ultimately cause coke formation and catalyst deactivation. To overcome this issue, the development of amorphous silica-alumina that involves a secondary mesoporous system to enhance the access of larger molecules to active sites has been investigated. In this study, a series of ASA under various conditions were prepared by a top-down method to investigate the formation mechanism. The synthesis was carried out by desilication of a commercial zeolite Y through a base treatment and the addition of a surfactant agent such as cetyltrimethylammonium bromide (CTAB). The prepared ASA showed enhanced surface area of more than 1000 m2/g and pore volume of 1.28 mL/g. The formation mechanism was investigated to understand how these ASAs are obtained while maintaining the order of the framework which has not been discovered yet. The ASA support was then used in final hydrocracking catalysts formulations to test its catalyst performance against other commercial ASAs. The physiochemical properties were studied using an array of multi characterization techniques. The N2 adsorption–desorption, pyridine adsorbed Fourier transform infrared (Py-FTIR) spectroscopy, 27Al magic angle spinning nuclear magnetic resonance (27Al MAS NMR), high-resolution transmission electron microscopy (HRTEM), Ammonia TPD (NH3-TPD) and so forth were employed to examine the physicochemical property of the ASA support.

Poster 15

Exploring Zeolite Potential for Hydrofluorocarbon Capture and Recycling: Insights from Molecular Simulations

Abrar A. Elhussien[a], Ismail Abdulazeez[b], Hassan Alasiri[a,c] and Wael A. Fouad[a,c,*]
[a] *Department of Chemical Engineering, King Fahd University of Petroleum and Minerals, Dhahran 31261, Saudi Arabia*
[b] *Interdisciplinary Research Center for Membranes and Water Security, King Fahd University of Petroleum and Minerals, Dhahran 31261, Saudi Arabia*
[c] *Interdisciplinary Research Center for Refining & Advanced Chemicals, King Fahd University of Petroleum and Minerals, Dhahran 31261, Saudi Arabia*

Hydrofluorocarbons (HFCs) play a vital role in refrigeration and air conditioning systems due to their exceptional thermal efficiency and properties. However, their high global warming potential (GWP) presents a significant environmental challenge, necessitating the development of innovative methods for their separation and recycling. This study evaluates the potential of zeolites as sustainable adsorbents for the effective capture, separation, and recycling of HFCs, focusing on methane derivatives (R32, R23, R41, R14) and ethane derivatives (R152a, R134a, R143a, R125). Advanced computational tools, including molecular dynamics (MD) and Grand Canonical Monte Carlo (GCMC) simulations, were employed to systematically analyze the adsorption behaviors of these HFCs in zeolites 13X, 5A, 4A, and 3A under a range of thermodynamic conditions. The findings revealed that zeolite 13X exhibits strong cation-fluorine interactions, making it particularly effective for adsorbing larger HFC molecules. Zeolite 5A, on the other hand, demonstrated a pronounced ability to engage in hydrogen bonding at low pressures and displayed entropic effects at higher pressures. Key insights include zeolite 5A's strong selectivity for R32 over R23 and zeolite 13X's exceptional adsorption capacity for R134a, driven by molecular size and polarity. Additionally, the study provided a detailed understanding of how framework characteristics, such as pore size and molecular interactions, govern the adsorption process, thereby elucidating critical structure-property-performance relationships. This research underscores the promise of zeolite-based materials as cost-effective, energy-efficient, and environmentally sustainable solutions for refrigerant management. By advancing our understanding of HFC separation and recycling, this work contributes to the development of green strategies for reducing the environmental impact of these potent greenhouse gases

Poster 16

A novel approach for the synthesis of carbon fiber from low-value heavy petroleum fractions

Ajay Sharma[1], Mustafa M. Amin[2] , Muhammad N. Siddiquee* [1,2]
[1] Interdisciplinary Research Center for Refining and Advanced Chemicals, King Fahd University of Petroleum & Minerals, Dhahran 31261, Saudi Arabia
2Department of Chemical Engineering, King Fahd University of Petroleum & Minerals, Dhahran, 31261, Saudi Arabia
**Email: muhammad.siddiquee@kfupm.edu.sa*

Carbon fiber's demand is increasing with a CAGR of 10.3% in aviation, defense, sports, and other sectors as a reinforcement material due to its lightweight nature, high tensile strength, and resistance to chemicals and temperature. Currently, 90% of carbon fibers (CFs) are polyacrylonitrile (PAN)-based where precursors cost consumes ~53% of the total production cost (11-260 $/kg CF) with 20-30 kgCO2eq/kg CF emissions, which limits the mass production and broader use of CF. In order to overcome these, the present research considered inexpensive carbon rich heavy petroleum fractions (e.g., heavy vacuum gas oil and vacuum residue). Petroleum derived asphaltenes is as good alternative of PAN for producing CF with a benefit of low cost (6.64 $/kg CF) and 68.7% lower CO2 emissions (Baritto et. al., Sustain. Mater. Technol. 36 (2023) e00627). In the reported literature, the inventors first used the pretreatment of these heavy petroleum fractions by autoxidation, or halogenation/dehalogenation methods followed by the solvent deasphalting to separate asphaltenes and used this as precursor material. But in our invention, we utilized a one-step process through an autoxidation pretreatment method by blowing air without mixing at a relatively lower temperature (up to 280°C). Blowing without mixing regulates free radical oxidation, enhancing carbon fiber precursor quality. The precursor presented enhancement in the desired characteristics as a CF precursor like enhanced viscosity, volatility, cross-linking, and softening point. Further, its derived green fiber also showed enhanced qualities such as uniformly smooth surface and no voids. By the use of this method, we synthesized the carbon fiber with a diameter of 40-50 μm and carbon content above 90 wt.% which is thermally stable (retained more than 85 wt.% at 850 °C). The use of air as an oxidizing agent, which is affordable, readily available, and useful for industrial application without the need for additional chemicals, can facilitate the production of carbon fiber from low-value heavy petroleum fractions.

Poster 17

Gen2 hydrocracking catalyst development for thermal crude to chemicals

Lianhui Ding[1], Mazin Tamimi[2], Faisal Alotaibi[1], Batool Altaher[1], Mohammad Bahhar[1], Faisal Mulla[1], Mohammad Aljishi[1]
[1]Catalyst Center of Excellence, R&D, Saudi Aramco, Dhahran, Saudi Arabia
[2]Fuel and Chemical Division, R&D, Saudi Aramco, Dhahran, Saudi Arabia

Abstract
Thermal Crude to Chemicals (TC2C™) technology economically optimizes the production of high-value chemicals from crude oil by tailoring the hydrogen content of selected crude oil components to make an optimum feed for a mixed feed steam cracker. TC2C™ is a large-scale integrated process that converts greater than 70 wt% of crude to chemicals. TC2C™ simplifies the crude conversion process compared to conventional crude oils to chemicals technologies while delivering higher chemicals yield. Saudi Aramco Technologies Company, Chevron Lummus Global (CLG), and Lummus Technology jointly developed this innovative and transformative process after several years of intense research at different scales. One of the important technologies for the TC2C process is novel hydrocracking catalyst to completely convert heavy oil to naphtha and lighter products which are preferred feedstocks for steam cracking. In order to further improve the olefin yield, reduce hydrogen consumption and catalyst synthesis cost, 2nd generation of hydrocracking catalyst was developed. The reaction performance of the catalyst was evaluated and compared with Gen1 catalyst DHC-23. By improving zeolite pore structures, adjusting the acidity, and optimizing hydrogenation and cracking functions of the catalyst, the iso-to-normal paraffin ratio, hydrogen consumption, and synthesis cost of Gen2 TC2C hydrocracking catalyst were reduced by 20-30%, 10%, and 10% respectively, which leads to more profit margin for refineries or petrochemical plants.

Poster 18

Hierarchical ZSM-11: A Catalyst for Converting Crude Oil into Light Olefins

Muhammad Waqas, Ziyauddin S. Qureshi*
Interdisciplinary Research Center for Refining and Advanced Chemicals
King Fahd University of Petroleum & Minerals, Dhahran-31261, Saudi Arabia
**Email: zqureshi@kfupm.edu.sa*

ZSM-11 is well-known as a shape-selective catalyst for various industrial processes; however, its potential in the direct conversion of crude oil to light olefins remains largely unexplored. In this study, we present a template-directed hydrothermal synthesis method to produce hierarchical ZSM-11 zeolites with tailored secondary mesopore sizes, acidities, and morphologies. The synthesis approach enabled the formation of uniform nanorods or aggregated nanosized zeolites with SiO_2/Al_2O_3 molar ratios of 30, 35, and 70.
The synthesized ZSM-11 catalysts, characterized by two-dimensional 10-membered ring straight channels, were utilized for the first time in the steam catalytic cracking of light crude oil to produce light olefins (C_2-C_4). Among the developed catalysts, ZSM-11(35) with an SiO_2/Al_2O_3 molar ratio of 35 exhibited superior performance, achieving a crude oil conversion of approximately 66.5% and a selectivity for C_2-C_4 light olefins of 44.1%. Compared to the diffusion challenges of sinusoidal channels in ZSM-5, the tortuous yet straight channels in ZSM-11 enhanced rapid mass transport while minimizing coke deposition, thus demonstrating its potential as an efficient catalyst for light olefin production.[1,2]

Poster 19

Co-Pyrolysis of Microalgae and Plastic Waste into Valuable Products

Fatima AlRadhi[1], Ahmad Nawaz[2], Shahina Riaz[2], Shaikh Abdur Razzak[1,2], Omar Abdelaziz[1,2] *
[1]Department of Chemical Engineering, King Fahd University of Petroleum & Minerals, Dhahran 31261, Saudi Arabia
[2] Interdisciplinary Research Center for Refining &Advanced Chemicals, King Fahd University of Petroleum & Minerals, Dhahran 31261, Saudi Arabia
**Email: omar.abdelaziz@kfupm.edu.sa*

The global demand for conventional energy sources is dramatically increasing, leading to the exploration of alternative, renewable energy options. Biomass, particularly microalgae, has emerged as a promising biofuel due to its abundance, low cost, sustainability, and rapid reproduction cycle. Simultaneously, the accumulation of plastic waste is a growing environmental concern. Chemical recycling is recognized as an effective treatment method, converting plastic waste into valuable chemicals. In the present study, the co-pyrolysis of Chlorella microalgae biomass and polypropylene (PP) plastic waste was investigated as a means of producing a sustainable and eco-friendly fuel. The objectives include studying the synergistic effects of co-pyrolysis at various feed ratios and temperatures, optimizing pyrolytic oil yield, and benchmarking the performance of the derived oil against fossil fuel alternatives. Preliminary thermogravimetric analysis revealed that plastic waste shows negligible weight loss at 470 °C, while Chlorella undergoes approximately 60% weight loss at this temperature. Initial yield calculations revealed a maximum bio-oil yield of 45% for pure plastic waste and a minimum of 20% for microalgae alone. Overall, this research underlines the feasibility of using co-pyrolysis to address both plastic waste disposal and sustainable energy production

Poster 20

Experimental Investigation of Co-Pyrolysis of Plastic Waste and Biomass: Biofuel Production and Comprehensive Characterization

Hayat Haddad[1], Ahmad Nawaz[2], Shaikh Abdul Razak[1, 2]
[1]*Department of Chemical Engineering, King Fahd University of Petroleum & Minerals, Dhahran 31261, Saudi Arabia*
[2] *Interdisciplinary Research Center for Refining &Advanced Chemicals, King Fahd University of Petroleum & Minerals, Dhahran 31261, Saudi Arabia*

The increasing global demand for sustainable energy and effective waste management necessitates innovative approaches to valorize biomass and plastic waste. Co-pyrolysis has emerged as a promising technique for producing biofuels and value-added products, addressing the dual challenges of renewable energy generation and environmental pollution. This study focuses on the experimental investigation of co-pyrolysis using biomass and plastic waste as feedstocks. The process was optimized to produce bio-oil, biochar, and syngas, with extensive characterization of the products conducted. Gas Chromatography-Mass Spectrometry (GC-MS) and Gas Chromatography-Flame Ionization Detector (GC-FID) were employed to analyze the chemical composition of the bio-oil, revealing the presence of hydrocarbons, oxygenated compounds, and aromatics suitable for fuel applications. Additionally, biochar was analyzed using Fourier Transform Infrared Spectroscopy (FTIR), Scanning Electron Microscopy (SEM), Brunauer– Emmett–Teller (BET) analysis, and Thermogravimetric Analysis (TGA) to determine its physicochemical properties and adsorption potential. The gaseous fraction was assessed for its calorific value and composition, indicating its suitability for energy recovery. This experimental work highlights the synergistic benefits of combining biomass and plastic waste for co-pyrolysis, providing a sustainable pathway for biofuel production and environmental waste mitigation. The study also evaluates the economic feasibility and scalability of this approach, offering insights into its practical implementation for renewable energy generation.

Poster 21

Conversion of polypropylene to chemicals in a reactive distillation system

Kagiso Bikane[a], Jingyang Bai[b], Marcos Millan[a,c]
[a] *Interdisciplinary Research Center for Refining and Advanced Chemicals, King Fahd University of Petroleum and Minerals, Dhahran 31261, Saudi Arabia*
[b] *Department of Chemical Engineering, Imperial College London, South Kensington Campus, London SW7 2AZ, UK*
[c] *Department of Chemical Engineering, King Fahd University of Petroleum & Minerals, Dhahran 31261, Saudi Arabia*

A considerable proportion of plastic waste generated is mismanaged and landfilled in the environment, posing ecological hazards and pollution. Pyrolysis is an attractive strategy to produce high-value gas and liquid products from plastic waste. However, due to the random C-C bond scission of plastic chains, achieving a narrow product distribution of specific high-value chemicals remains a great challenge with reported products in literature ranging from carbon numbers of C_5 to C_{60}. This hinders downstream processes aiming for conversion towards specific chemicals. In this work, PP was pyrolyzed in a novel reactive distillation system employing a vertical partial condenser acting as a partial reflux, returning heavy hydrocarbon products (> boiling point of 60 °C) to the reactor, targeting simultaneous production of light olefins and monoaromatics. Light liquid products that escaped the reactor system were condensed while gases were passed to an online GC/MS system. An HZSM-5 catalyst was used for catalytic pyrolysis at a temperature of 400 °C and reaction time of 90 min. A comparison of the catalytic pyrolysis of PP with and without a partial reflux system is shown in Figure 1. The reflux system exhibited a significant shift towards light products with a higher gas yield. The liquid product had a higher composition of the high-value BTEX using partial reflux (45 wt.%) compared to a system without reflux (26 wt.%). A maximum carbon number of C_{12}, accounting for 95 wt.% of the pyrolysis products with the rest being solid residue, was obtained using partial reflux while its absence led to a broad gas and liquid distribution of C_5 – C_{38} (Figure 1c). This demonstrates the superior control of the reaction products owing to further cracking in a reactive distillation system. Gas product characterisation showed high concentrations of H_2 and light olefins (mainly propene, butene and ethylene) from the catalytic pyrolysis of PP (Figure 1d). Amongst the limited pyrolysis research on reactive distillation, this study is the first to report aromatics, which are obtained alongside other high-value chemicals including hydrogen and gas and liquid olefins.

Poster 22

Ensemble Technique for Predicting Dehydrogenation Products Yield during n-Butane Oxidative Dehydrogenation

Gazali Tanimu[1,2], Jimoh O. Ajadi[1,3], Hassan Alasiri[1,2]
[1]*Interdisciplinary Research Center for Refining & Advanced Chemicals, King Fahd University of Petroleum & Minerals, Dhahran, Saudi Arabia.*
[2]*Department of Chemical Engineering, King Fahd University of Petroleum & Minerals, Dhahran, Saudi Arabia.*
[3]*Department of Mathematics, King Fahd University of Petroleum & Minerals, Dhahran, Saudi Arabia.*
**Email: gazali.tanimu@kfupm.edu.sa*

In this study, we leverage the power of machine learning to predict dehydrogenation (DH) product yield and unravel the factors influencing the product distribution. A comprehensive dataset obtained from experiments conducted in a fixed-bed reactor under varying temperatures, feed ratios (O_2/n-butane), and metal oxide loadings (Ni, Fe, Co, Bi, Mo, W, Zn, and Mn) on an aluminum oxide support served as the basis for model development. Three supervised machine learning models, Boosted Tree (BT), Extreme Gradient Boosting Linear (XGBL), and Support Vector Machine Radial (SVMR), were evaluated. The ensemble technique of the three models showed remarkable accuracy, with an RMSE of 1.65 and MAE of 1.14 on the test dataset, and it demonstrated robust generalization capabilities by capturing 87% of the variation in DH yield. In the feature importance analysis of the selected models, Ni, Mo, Co, and W emerged as critical factors influencing the DH yield. The ability of the ensemble model to predict DH yields opens new avenues for optimizing DH products and designing more advanced catalysts.

Poster 23

Rational Design of N-Aryl PNP Ligands for Controlling Product Profiles in Cr-Catalyzed Ethylene Oligomerization

Abdulrahman Musa[1], E. A. Jaseer[1,2*], Samir Barman[1], Nestor Garcia[1], Mohamed Elanany[3], Motaz Khawaji[3]
[1] Interdisciplinary Research Center for Refining & Advanced Chemicals, King Fahd University of Petroleum & Minerals, Dhahran 31261, Saudi Arabia
[2] Department of Chemistry, King Fahd University of Petroleum & Minerals, Dhahran 31261, Saudi Arabia
[3] Aramco R&D Division, King Abdullah University of Science and Technology, Thuwal 23955, Saudi Arabia
**Email: jaseer@kfupm.edu.sa*

A systematic investigation was carried out to evaluate the influence of fused polyaromatic N-aryl substitutions in PNP-type ligands on chromium-catalyzed ethylene tetramerization reactions. The study explored the incorporation of varying steric profiles around the ligand backbone to influence catalytic pathways and to achieve targeted 1-C_8 and 1-C_6 olefin products. Enhanced C_8 olefin selectivity (up to 74.6 wt% with a C_8/C_6 ratio of 3.4) was achieved when the N-aryl connection to the aromatic unit was placed at the 2-position. In contrast, higher C_6 olefin selectivity (up to 33.7 wt% with a C_8/C_6 ratio of 1.9) was observed when the PNP ligand was anchored at the 1- or 6-position. Detailed catalytic studies revealed that fine-tuning the steric bulk at the N-site, by strategically selecting N-aryl moieties of varying sizes, provides a rational approach to modulate the selectivity of desired olefin products.

Poster 24

Mesoporosity Engineering in Zeolites for Enhanced Catalytic Performance: Post-Synthesis Modification and Adsorption Study

Nida Tasneem[a,b], Hassan-Alasiri [a,b*], Shakeel Ahmed[b] ,
[a] *Department of Chemical Engineering, King Fahd University of Petroleum & Minerals, Saudi Arabia*
[b] *Interdisciplinary Research Center for Refining and Advanced Chemicals, King Fahd University of Petroleum & Minerals, Saudi Arabia*
**Email: g202216060@kfupm.edu.sa, nidatasneem642017@gmail.com*

Zeolites are now widely used as a main heterogeneous catalyst in many chemical processes. But their diffusion limits lead to ineffective catalyst use, which in turn causes early coking and deactivation. It has been determined that hierarchical zeolites represent a promising advanced material capable of overcoming the limitations of microporous zeolites in catalytic processes and improving the physical transport of larger reactant molecules. This work provides new understanding of the surfactant-templating mechanism that produces tailored intracrystalline mesoporosity in various zeolites, such as ZSM-5 and Y zeolite while preserving its crystalline structure and essential properties. To optimize pore architectures and preserve the inherent properties of the zeolites, the degree of mesoporosity was adjusted by the use of controlled dealumination and desilication procedures. To examine the structural and chemical changes brought about by the modification process, characterization methods including nitrogen physisorption (BET), scanning electron microscopy (SEM), transmission electron microscopy (TEM), X-ray diffraction (XRD), and Temperature-programmed desorption (TPD) were used. Furthermore, using molecular studio modeling tools, adsorption isotherms were produced that shed light on the modified zeolite materials' adsorption behavior. Software-generated isotherms and experimental data are compared, which provides additional evidence of the modification process' efficacy. Through the creation of innovative zeolite materials with tailored porosity and enhanced adsorption capabilities, this integrated approach advances the field and offers promising applications in catalysis, gas separation, and other areas.

Poster 25

A Lumped Kinetic Model for Hydro-Dearylation Process

Qi Xu, Haitao Huang, Mohammad Rebh
R&D Center, Saudi Aramco, Dhahran, Saudi Arabia

The future of refining and chemicals relies on sustainable growth of petrochemicals, which has accelerated through the on-going energy transition in the refining industry from fuels to chemicals. In a typical aromatics recovery complex, BTX (benzene, toluene, and xylenes) are main products while a reject heavy aromatics stream contains valuable aromatics in the form of alkyl-bridged non-condensed multi-aromatics. This relatively low value reject stream has limited sustainability due to its nature of stream composition. To overcome this limitation, we have developed a novel catalytical technology to recover valuable monoaromatics from this reject stream. This process has been proven conceptually by experiments conducted in a pilot plant testing unit, and the process conditions have been preliminarily optimized.

Furthermore, a lumped kinetic model has been developed to better understand the reaction mechanisms of the process. This model has been validated by the extensive pilot plant testing. A reaction network is built with six pseudo-components, each representing species with similar characteristics. The reaction pathways contain equilibrium reactions and hydrocracking reactions. They illustrate how the multi-ring aromatics are converted into the mono-ring aromatics. The reaction rates of constant and activation energies are obtained from experimental results. The predictions from the kinetic model are compared with the experimental values and good agreement is obtained.

In conclusion, a lumped kinetic model has been developed to align with the newly developed catalytic process, offering valuable insights into its operational mechanisms. This study has enabled exploration of various scenarios to optimize the operation of the entire aromatics recovery complex and enhanced the sustainability of this technology for the future.

Poster 26

Development of Tin-Containing Mesoporous Silica Catalysts for Baeyer-Villager Oxidation of Cyclic Ketones

Mashail Al-Qahtani[1], Selvaraj Manickam[2], Emad Shafi[1], Mohammad Aljishi[1], Rawan Alawami[1], Sakinah Bin Hamad[1], Ahmad Alaseel[3], Hatim Dafalla[4], Chang Sik Ha[5]
[1] *Saudi Aramco Research & Development Centre*
[2] *Department of Chemistry, Faculty of Science, King Khalid University*
[3] *Research and Analytical Service Department*
[4] *Core Research Facilities, King Fahd University of Petroleum & Minerals*
[5] *Department of Polymer Science & Engineering, Pusan National University*

In this study, the mesoporous SnKIT-6 catalysts with different n_{Si}/n_{Sn} ratios (2, 6 and 12) were synthesized by pH-aDH method with and without presence of HCl. The prepared samples underwent detailed characterization to investigate their physicochemical properties. Characteristic results confirm that the SnKIT-6 catalysts have a highly mesoporous nature and have huge amounts of Sn^{4+} coordinated tetrahedrally on the surface of silica walls. The mesoporous SnKIT-6 catalysts were used in Baeyer Villiger (BV) Oxidation of cyclic ketones. The catalytic results were collected using different catalytic optimal conditions. The SnKIT-6(2) with n_{Si}/n_{Sn} ratio of (2) synthesized in absence of HCl exhibited the best performance with a cyclohexanone (CH=O) conversion of 91% and a ε-caprolactone (CL=O) selectivity of 98%. However, it further produced 100% CL=O selectivity with 95% CH=O conversion and leads to almost 100% selectivity of lactones from the corresponding cyclic ketones when recycled at run number four.

Poster 27

Efficient Synthesis of Dimethyl Carbonate from CO_2 using Cerium Oxide-Based Catalysts

Niladri Maity, Samiyah A. Al-Jendan and Norah Fayez
Interdisciplinary Research Center for Refining and Advanced Chemicals
King Fahd University of Petroleum & Minerals, Dhahran-31261, Saudi Arabia

The increasing concentration of carbon dioxide (CO_2) in the Earth's atmosphere presents a pressing environmental challenge, necessitating innovative solutions to mitigate climate change. While CO_2 utilization has significant potential, only a few processes have been developed to effectively convert it into valuable industrial chemicals, such as urea, salicylic acid, and polycarbonate polyol. Among these, dimethyl carbonate (DMC) stands out, with a global demand driven by its diverse applications, including its use as a feedstock for the production of polycarbonate, an electrolyte in lithium-ion batteries, and a chemical intermediate for polyurethanes. Additionally, DMC is recognized as a green solvent due to its low toxicity, biodegradability, and water miscibility.

Various large-scale industrial processes for DMC production, such as the phosgene route, transesterification, and liquid-phase oxidative carbonylation of methanol, are currently employed. However, these methods have significant limitations, including the use of hazardous phosgene, high production costs, and safety risks such as explosion hazards. In contrast, the direct synthesis of DMC from CO_2 and methanol (CH_3OH) offers a greener and safer alternative due to the non-toxic nature of the starting materials and the sustainability of the process (Scheme 1).

Scheme 1. Catalytic conversion of CO_2 into dimethyl carbonate.

Catalyst systems for this process include ionic liquids, alkali carbonates, heteropolyacids, supported metal nanoparticles, and metal oxides. Among these, cerium oxide (CeO_2)-based materials stand out for their stability, CO_2 activation efficiency, and promotion of methanol carboxylation to DMC [1]. Low DMC yields can be improved with an effective dehydrating agent.

A series of cerium oxide (CeO_2) nanorod catalysts doped with sulfur (S) and nitrogen (N) were synthesized and thoroughly characterized using Raman, PXRD, XPS, TEM, N_2-physisorption, CO_2, and NH_3 TPD [2]. The superior efficiency of the heteroatom-assisted CeO_2 catalysts towards DMC yield from CO_2 and methanol could be attributed to their improved surface acidity, basicity, and increased oxygen vacancies compared to undoped CeO_2 nanomaterials.

Poster 28

Single-step electrochemical conversion of CO_2 to C_4 products with Cu based catalyst unveiling mechanisms through DFT and In Situ Raman spectroscopy

Esraa Kotoba , Khalid Alhooshania , [b] , Saheed A. Ganiyu[a,b] *
[a] *Department of Chemistry, King Fahd University of Petroleum and Minerals, Dhahran 31261, Saudi Arabia.*
[b] *Interdisciplinary Research Center for Refining & Advanced Chemicals, King Fahd University of Petroleum and Minerals*

The market value of C4 products, such as n-Butanol and Butene, is substantial, with 1 tons valued at approximately $1,300, highlighting the significant economic potential of this conversion process. Achieving direct CO2 conversion to C4 products in a single step, especially at high Faradaic efficiencies, presents a formidable challenge. Traditional methods often rely on intricate cascade systems to attain the necessary selectivity and efficiency. In this study, we successfully demonstrated a breakthrough in electrochemical CO2 conversion to C4 products without the need for a cascade system. By utilizing a Cu-based morphology catalyst, we achieved notably high Faradaic efficiency for C4 products and ethylene at industrial current densities, while maintaining exceptional stability throughout the process. Additionally, the underlying reaction mechanisms was further elucidated through Density Functional Theory (DFT) calculations and operando Raman spectroscopy, offering deeper insights into the catalytic behavior. This work presents a highly promising and sustainable single-step process for large-scale C4 products, with far-reaching implications for both industrial applications and environmental sustainability

Poster 29

Synergistic Role of La and Ni Species on CeO_2 decorated SiO_2 Catalyst for Enhanced CO_2 Conversion to Synthetic Natural Gas (SNG)

Ijaz Hussain [1], Saheed A. Ganiyu [1,2], Khalid Alhooshani [1,2*]
[1] Interdisciplinary Research Center for Refining & Advanced Chemicals, King Fahd University of Petroleum & Minerals, Dhahran, 31261, Saudi Arabia
[2] Department of Chemistry, King Fahd University of Petroleum and Minerals, Dhahran, 31261, Saudi Arabia
**Email: hooshani@kfupm.edu.sa.*

The consumption of global energy derived from fossil fuels has increased tremendously, which has caused climate change due to extensive CO_2 emissions. CO_2 methanation is one of the top-notch approaches to mitigate CO_2 emissions by converting them into value-added synthetic natural gas (SNG). This study focuses on the synthesis of Lanthanum (La) promoted Ni-CeO_2-SiO_2 catalyst, synthesized by microemulsion method to investigate the CO_2 methanation. The addition of La considerably improved the catalytic activity of the Ni-CeO_2-SiO_2 catalyst, resulting in more effective CO_2 methanation. This improvement is ascribed to accessible basic sites on the metal-support interface of the fibrous La-Ni@CeO_2 catalyst, resulting in excellent CO_2 conversion (84%), selectivity (96%), and yield (81%) of methane. Notably, the fibrous La-Ni@CeO_2 catalyst exhibits remarkable thermal stability over 50 hours, contrasting with the deactivation observed in other catalysts. The exceptional performance of the La-Ni@CeO_2 is attributed to the energetically favorable decomposition of CO_2 * to CO, rather than hydrogenation to HCOO. This study provides insights into the complex interplay among oxygen vacancies, basic sites, and reaction pathways in CO_2 methanation over La-Ni@CeO_2-SiO_2, elucidating their potential role for sustainable methane production from CO_2 conversion.

Poster 30

Carbon Nitride Modified Highly Reduced Graphene Composite as a Potential Material for CO_2 Utilization: Insights from First-Principle DFT Calculations

Abdulraheem K. Bello[a], Abdulaziz A. Al-Saadi[a,b]
[a] *Department of Chemistry, King Fahd University of Petroleum & Minerals, Dhahran 31261, Saudi Arabia*
[b] *Interdisciplinary Research Center for Refining and Advanced Chemicals, King Fahd University of Petroleum & Minerals, Dhahran 31261, Saudi Arabia*

The level of carbon dioxide in the atmosphere continues to increase as a result of anthropogenic activities which impose greater threats in terms of global warming and climate change. Catalytic reduction reactions of CO2 (CO2RR) leading to useful products are considered an attractive means for carbon utilization and greenhouse gas mitigation. Graphitic carbon nitride photocatalysts have attracted a great deal of interest due to their low band gap energy and good absorption ability, owing to a more extended conjugated network and the participation of the lone pair on the N atom with the π-conjugated system of heptazine motifs. However, such two-dimensional materials still need to be upgraded in regard to the photo-generated carrier dynamics for a more efficient photoactivity. Z-scheme hetero-junctions constructed by two or more suitable semiconductors can not only promote the spatial separation of photoinduced charge carriers, but also retain relatively stronger oxidative and reductive ability. Herein, we have explored the catalytic activity of g-C3N5 nanosheets combined with highly reduced graphene (HRG) towards the CO2 reduction on the basis of the density functional theory (DFT). In order to gain insights into the appropriateness and capability of such nanomaterials, key electronic and optical properties, such band structures, work functions, frontier molecular orbitals, and UV-Vis absorption, were theoretically derived and thoroughly analyzed. The band gap is predicted to drop from 1.8 to less than 0.5 eV upon combining HRG with g-C3N5, showing the ease of separation of photogenerated electron/hole pairs as a heterojunction is formed. The magnitude of the work function of pristine g-C3N5 (5.4 eV) decreases to 5.2 eV when HRG is incorporated, which is indicative of a higher tendency of the new material to have a better reduction capability. The simulated electronic spectra depict a shift of λmax from 360 to 540 nm, before and after the heterojunction modification, respectively. The change in the absorption pattern is evidence of the increase in the photocatalytic performance of the nanocomposite materials. Possible pathways of CO2RR to C1 products using the designed g-C3N5/HRG surface were computed and assessed. A limiting potential of -0.16 V leading to CH4 final product was calculated. We believe that these results provide valuable insights for rational design of effective photocatalysts for CO2RR applications.

Technology in Fuels & Petrochemicals

Advancing Sustainable Solutions

Venue: KFUPM Dhahran Techno Valley, Innovation Cluster Bldg., Auditorium 1-111

February 10-11, 2025

Day One: Monday, February 10, 2025

OPENING REMARKS SESSION — *Chairman: Dr. Wael Fouad, KFUPM*

8:30 *Opening Remarks*

- **Dr. Muhammad Al-Saggaf**, *President, KFUPM*
- **Mr. Hitoshi Masuda**, *Chief Technology Officer, Senior Executive Director, JCCP*
- **Dr. Faisal Al-Otaibi**, *Director, R&D Center, Saudi Aramco*
- **Dr. Naonobu Katada**, *Vice President, JPI*

SESSION ONE WASTE-TO-CHEMICALS — *Chairman: Dr. Takashi Toyao, Hokkaido Univ.*

9:00 **1.** Keynote: **Waste to sustainable aromatics, olefins and aviation fuels by zeolite-based catalysis**; *Prof. Konstantinos Triantafyllidis, CHEM, CRAC, KFUPM*

9:30 **2.** **Catalytic transformation of LDPE into fuel oil enriched with aromatic hydrocarbons**; *Dr. Muhammad Akthar, CRAC, KFUPM*

9:50 **3.** **Pyrolysis oil upgrading to aromatics**; *Dr. Miao Sun, Saudi Aramco*

10:10 *Coffee break*

SESSION TWO SUSTAINABLE SOLUTIONS — *Chairman: Dr. Shakeel Ahmed, KFUPM*

10:40 **4.** **Recycling of waste plastics without consumption of other materials: Shape selective cracking of polyolefin on H-MFI zeolite catalysts with recovery of cyclooctane solvent**; *Prof. Naonobu Katada, Tottori University*

11:00 **5.** **Kinetic influence of Ba on cobalt-catalyzed ammonia decomposition reaction**; *Dr. Zahra Almisbaa, Saudi Aramco*

11:20 **6.** **Sustainable passive solutions for reduced energy use in cooling applications**; *Dr. Issam Gereige, Saudi Aramco*

11:40 *Prayer & Lunch Break*

SESSION THREE NEW CHALLENGES: PROCESSES/FUELS — *Chairman: Dr. Faisal Alshafei, Saudi Aramco*

13:00 **7.** **Sustainability without borders: Advancing partnerships between the Kingdom of Saudi Arabia and Japan**; *Eng. Zainab Nasif and Ms. Ahaad Turki, OSP*

13:30 **8.** **North American market research: Energy transition to renewables from policies and technologies**; *Dr. Satoshi Ogano, JPI*

13:50 **9.** **Ultrasonically enhanced oxidative desulfurization of heavy fuel oil: Pilot scale test results and techno-economics assessment**; *Dr. Shekhar Kulkarni, KAUST*

14:10 **10.** **Catalysis reinvented: Sustainable materials and ultrafast laser insights for economic and environmental advancement**; *Dr. Ahmed El-Zohry; CIPR, KFUPM*

14:35 - 15:50 *Poster Session One* *(See Page 3)*

aramco

Technology in Fuels & Petrochemicals

Advancing Sustainable Solutions

Day Two: Tuesday, February 11, 2025

SESSION FOUR ADVANCED CATALYST TECHNOLOGIES *Chairman: Dr. Naonobu Katada, JPI*

08:30 11. Keynote: **Accelerated discovery of heterogeneous catalysts using machine learning**; *Dr. Takashi Toyao, Hokkaido University*

09:00 12. **Mechanochemical route to synthesize zeolites as novel method and their characterization with XAFS**; *Emeritus Prof. Atsushi Muramatsu, Tohoku University*

09:20 13. **Catalytic cracking technology for VGO application**; *Ms. Maryam Taher, Saudi Aramco*

09:40 ***Coffee Break***

SESSION FIVE SUSTAINABLE CHEMICALS *Chairman: Dr. Ziyauddin Qureshi, KFUPM*

10:10 14. Keynote: **Rational design of heterogeneous catalyst for the production of sustainable chemicals**; *Dr. Javier Martinez, KAUST*

10:40 15. **Synergistic effects of S-C_3N_4 on cobalt ferrite catalysts for oxygen evolution reaction and enhanced electrochemical conversion of biomass derived 5-hydroxy-methylfurfural into 2,5-furandicarboxylic acid**; *Dr. Wasif Farooq, CHE, KFUPM*

11:00 16. **Tuning the properties of ZSM-5 additive for co-cracking of waste plastics dissolved in vacuum gas oil**; *Dr. Abdulkadir Tanimu, CRAC, KFUPM*

11:20 17. **Catalytic microwave assisted pyrolysis of waste plastics**; *Dr. Abdul Gani Abdul Jameel, CHE, KFUPM*

11:40 ***Prayer & Lunch Break***

13:00 - 14:15 ***Poster Session Two (See Page 4)***

SESSION SIX CO_2 CAPTUR AND CONVERSION *Chairman: Dr. Rashed Aleisa, Saudi Aramco*

14:20 18. Keynote: **Hydrogenation of CO_2 for carbon neutrality: Enhancing efficiency with membrane reactors**; *Prof. Masahiko Matsukata, Waseda University*

14:50 19. **Steering electrochemical CO_2 reduction toward liquid fuels and chemicals**; *Mr. Husain Naji, Saudi Aramco*

15:10 20. **Development of highly dispersed Pd nanoparticles supported on dendritic mesoporous heterogeneous catalyst for efficient CO_2 hydrogenation to methanol**; *Dr. Mohnnad Alabsi, Saudi Aramco*

15:30 ***Closing Remarks, Symposium Ends***
Dr. Wael Fouad, CRAC, KFUPM

Each presentation includes 5-minutes Q&A

Technology in Fuels & Petrochemicals

Advancing Sustainable Solutions

POSTER SESSION ONE: MONDAY, FEBRUARY 10, 2025 ***Coordinator: Dr. Jaseer EA, KFUPM***

14:35	P1.	**Post-synthesis functionalization of covalent organic framework for CO_2 capture from DAC**; *Mona Al-Otaibi, Saudi Aramco*
14:40	P2.	**DFT study of CO_2 hydrogenation over alumina-based catalysts via RWGS reaction**; *Abdulrahman Yassin, Saudi Aramco*
14:45	P3.	**Enhancing hydrogen production: Acidic-basic structural modification of nickel-based catalysts for ammonia decomposition**; *Reem Albashrawi, Saudi Aramco*
14:50	P4.	**Electrochemical ammonia decomposition over nickel and cobalt nanostructures on nickel foam for sustainable hydrogen production**; *Umar Jafar, Mustapha Umar, Yahaya Gambo, Khalid Alhooshani, Yousef Alsunni, CHEM, CHE, CRAC, KFUPM*
14:55	P5.	**Maximizing CO production from CO_2 conversion for e-fuels: Reverse water-gas shift reaction**; *Emad Al-Shafei, Saudi Aramco*
15:00	P6.	**Quantitative risk assessment of hydrogen refueling stations: A case study for Saudi Arabia**; *Bashir Hashim, Sunhwa Park, CHE, CRAC, KFUPM*
15:05	P7.	**Selecting hydrogen storage in Saudi Arabia: A multi-criteria framework integrating FAHP for uncertainty and safety**; *YooJeong Oh, Hans Pasman, Safyan Khan, Sunhwa Park, CHE, CRAC, CHTC, KFUPM, Texas A&M USA*
15:10	P8.	**Exploring the synergy between CoP and Co_2P phases: A comparative study of cobalt phosphides phases and their composite for hydrogen evolution reaction**; *Wajiha Fatima, Tarek Kandiel, CHEM, CHTC, KFUPM*
15:15	P9.	**Decorating $Zr_{12}O_{12}$ nanocage with transition metals (Ti, V, Cr, or Mn) to act as a single-atom catalyst for water splitting application**; *Sajjad Hussain, Abdulaziz Al-Saadi, CHEM, CRAC, KFUPM*
15:20	P10.	**Production of hydrogen and value-added chemicals through microwave assisted pyrolysis (MAP) of Arabian Crude Oils**; *Intisar Ul Hassan, Aniz Ummer, Abdul Gani Abdul Jameel, CHE, CRAC, KFUPM*
15:25	P11.	**Dry reforming of methane to syngas using Ni/Al_2O_3 catalysts promoted by lanthanides**; *Imtiaz Ul Hasan, Zuhair Malaibari, Shakeel Ahmed, CHE, CRAC, KFUPM*
15:30	P12.	**Catalytic performance in dry reforming of methane: A comparative study of SiO_2 encapsulated and unencapsulated Ni/Cao-Al_2O_3 catalysts**; *Sherif Alabi, Sagir Adamu, Abdallah Al-Shammari, Mohammad Hossain, CHE, CRAC, KFUPM*
15:35	P13.	**Synthesis of ZSM-nonorods for low temperature CO_2 reforming of methane to syngas**; *Yahuza Abdulai, Ijaz Hussain, Khalid Alhooshani, Saheed Ganiyu, CHEM, CRAC, KFUPM*
15:40	P14.	**Case study on the acid sites formation and morphology in amorphous silica-alumina**; *Batool Altaher, Saudi Aramco*
15:45	P15.	**Exploring zeolite potential for hydrofluorocarbon capture and recycling: Insights from molecular simulations**; *Abrar Elhussien, Ismail Abdulazeez, Hassan Alasiri, Wael Fouad, CHE, CMWS, CRAC, KFUPM*

Technology in Fuels & Petrochemicals

Advancing Sustainable Solutions

POSTER SESSION TWO: TUESDAY, FEBRUARY 11, 2025 ***Coordinator: Dr. Jaseer EA, KFUPM***

13:00 P16. **A novel approach for the synthesis of carbon fiber from low-value heavy petroleum fractions**; *Ajay Sharma, Mustafa Amin, Muhammad Siddiquee, CHE, CRAC, KFUPM*

13:05 P17. **Gen2 hydrocracking catalyst development for thermal crude to chemicals**; *Lianhui Ding, Saudi Aramco*

13:10 P18. **Hierarchical ZSM-11: A catalyst for converting crude oil into light olefins**; *Muhammad Waqas, Ziyauddin Qureshi, CRAC, KFUPM*

13:15 P19. **Co-pyrolysis of microalgae and plastic waste into valuable products**; *Fatima AlRadhi, Ahmad Nawaz, Shahina Riaz, Shaikh Abdur Razzak, Omar Abdelaziz, CHE, CRAC, KFUPM*

13:20 P20. **Experimental investigation of co-pyrolysis of plastic waste and biomass: Biofuel production and comprehensive characterization**; *Hayat Haddad, Ahmad Nawaz, Shaikh Abdul Razak, CHE, CRAC, KFUPM*

13:25 P21. **Conversion of polypropylene to chemicals in a reactive distillation system**; *Kagiso Bikane, Jingyang Bai, Marcos Millan, CHE, CRAC, KFUPM, ICL, UK*

13:30 P22. **Ensemble technique for predicting dehydrogenation products yield during *n*-butane oxidative dehydrogenation**; *Gazali Tanimu, Jimoh Ajadi, Hassan Alasiri; CRAC, CHE, MATH, KFUPM*

13:35 P23. **Rational design of N-aryl PNP ligands for controlling product profiles in Cr-catalyzed ethylene oligomerization**; *Abdulrahman Musa, E. A. Jaseer, Samir Barman, Nestor Garcia, Mohamed Elanany, Motaz Khawaji, CRAC, CHEM, KFUPM, Saudi Aramco*

13:40 P24. **Mesoporosity engineering in zeolites for enhanced catalytic performance: Post-synthesis modification and adsorption study**; *Nida Tasneem, Hassan Alasiri, Shakeel Ahmed, CHE, CRAC, KFUPM*

13:45 P25. **A lumped kinetic model for hydro-dearylation process**; *Qi Xu, Saudi Aramco*

13:50 P26. **Development of tin-containing mesoporous silica catalysts for Baeyer-Villager oxidation of cyclic ketones**; *Mashail Al-Qahtani, Saudi Aramco*

13:55 P27. **Efficient synthesis of dimethyl carbonate from CO_2 using cerium oxide-based catalysts**; *Niladri Maity, CRAC, KFUPM*

14:00 P28. **Single-step electrochemical conversion of CO_2 to C_4 products with Cu based catalyst unveiling mechanisms through DFT and in situ Raman spectroscopy**; *Esraa Kotob, Khalid Alhooshani, Saheed Ganiyu, CHEM, CRAC, KFUPM*

14:05 P29. **Synergistic role of La and Ni species on CeO_2 decorated SiO_2 catalyst for enhanced CO_2 conversion to synthetic natural gas (SNG)**; *Ijaz Hussain, Saheed Ganiyu, Khalid Alhooshani, CHEM, CRAC, KFUPM*

14:10 P30. **Carbon nitride modified highly reduced graphene composite as a potential material for CO_2 utilization: insights from first-principle DFT calculations**; *Abdulraheem Bello, Abdulaziz Al-Saadi, CHEM, CRAC, KFUPM*